Sebastian Bücker, Marcus Düwell, Andreas Kaminski, Michael Leyer (eds.)
Trust, Responsibility, and Digital Governance

Digital Society | 79

Sebastian Bücker works as a scientific researcher at the Institute of Philosophy, TU Darmstadt. Using both his backgrounds in computer science (M.Sc.) and philosophy of technology (M.A.), his research focusses on how current developments in AI can be connected to philosophy, especially regarding the ascription of capabilities associated with autonomy.

Marcus Düwell is a professor of philosophy at the Institute for Philosophy at Technische Universität Darmstadt. His research interests include foundational questions of moral and political philosophy, philosophical anthropology, and applied ethics.

Andreas Kaminski is a professor of philosophy of science and technology at Technische Universität Darmstadt. His research areas include social epistemology (philosophy of trust and testimony), technical epistemology (the role of technology in science) and politics of technology. He is a senior scientist at the High-Performance Computing Center of the Universität Stuttgart (HLRS).

Michael Leyer (PhD) is a professor of business administration and chair of Digitalization and Process Management at Universität Marburg. He is also an adjunct professor at Queensland University of Technology. The main focus of his research is the sociotechnical consideration of future technologies in organizations.

Sebastian Bücker, Marcus Düwell, Andreas Kaminski, Michael Leyer (eds.)

Trust, Responsibility, and Digital Governance

Regulation of AI and Blockchain Technology from a Capacity-Based Perspective

[transcript]

This publication was made possible by the Open Access Monographs Fund of the University and State Library Darmstadt.
Open Access funding provided by the Open Access Publishing Fund of Philipps-Universität Marburg.

Bibliographic information published by the Deutsche Nationalbibliothek
The Deutsche Nationalbibliothek lists this publication in the Deutsche Nationalbibliografie; detailed bibliographic data are available online at https://dnb.dnb.de

transcript Verlag | Hermannstraße 26 | D-33602 Bielefeld | live@transcript-verlag.de

Cover design: Kordula Röckenhaus
Printing: Elanders Waiblingen GmbH, Waiblingen
https://doi.org/10.14361/9783839410974
Print-ISBN: 978-3-8376-7802-4 | PDF-ISBN: 978-3-8394-1097-4
ISSN of series: 2702-8852 | eISSN of series: 2702-8860

Printed on permanent acid-free text paper.

Contents

Acknowledgments

The present volume is the result of a research group that was funded by the Hessian *Centre Responsible Digitality* (https://zevedi.de), which we would like to thank for providing the environment that enabled the stimulating discussions of our interdisciplinary group. This environment made it possible for us to share and discuss our different perspectives on the problems of digital governance, allowing us to develop new insights that are expressed in this volume. We also want to thank the Open-Access funds of the University and State Library in Darmstadt and of Marburg University, which by generously funding this publication have allowed us to share our results and ideas with the scientific community. Further, we would like to thank our publisher *transcript* and in particular Jonas Geske for accompanying the publishing process, especially through his swift and informative answers to all our questions. Finally, we want to thank Melina Licht and Thorid Charlotte Schäfer, who meticulously proofread this volume and whose attention to detail harmonized all its texts.

Introduction: Trust, Responsibility, and Digital Governance

Sebastian Bücker, Marcus Düwell, Andreas Kaminski, Michael Leyer

Digital tools are increasingly influencing and shaping more aspects of our lives. We communicate via information technology, we develop a shared understanding by reading shared texts, watching pictures and videos, we make decisions based on information and recommendations provided by digital systems. When we are in contact with professionals, like physicians, financial advisers, lecturers, or travel agents, they often communicate with us based on information they obtain from digital tools; often, this communication itself is shaped by digital systems. More and more digital systems are supposed to implement functional roles that were previously performed by humans, and even our social and personal relationships are often initiated, mediated, and shaped by digital tools.

This ubiquity of digital tools comes with specific challenges. The question of whether we can trust the information we receive, the institutions we have to deal with and to what extent we can trust each other, depends to a large extent on the way digital tools are shaped, how they function, and how they are controlled. The physician searching in a computer for medicine for an illness is not in a position to judge how the program is influenced by the pharmaceutical industry. The student who asks ChatGPT about the subject for his homework cannot judge to which extent the given information can be taken as testimony. The teen who shares photos from the last vacation on social media cannot judge who has access to this data and what conclusions are drawn from it. We all are dependent on the governance of these digital aids and our trust in the social world and our trust in each other depends on the quality of this digital governance.

But how can we tell whether that is actually the case? This question is often taken to mean that we should develop a set of criteria and then evaluate digital systems by checking whether they meet those criteria. Criteria that are often proposed in these contexts are, for example, transparency, non-discrimination, non-abusiveness, etc. Next to the question of whether there is consensus about what the criteria should be, first there are some questions that are much more basic and tacitly presupposed in debates on the normative assessment of digital systems, namely questions regarding our ability to understand the digital systems in the first place. Before we can ask whether a digital system is designed in a non-discriminatory way, we have to ask: Are we even able to understand how digital systems are designed? Do we possess the capacity to understand how they shape our actions, our communications, and our decisions? If we are not able to understand how digital systems work, the ethical and legal assessment of whether they meet normative criteria cannot even be posed. And if this more basic capacity of understanding and evaluating is at stake, we can furthermore ask: How can this capacity to assess systems and their effects be cultivated and developed? How can digital governance support individuals and institutions in building such capacities? Thus, this edition asks those questions that are necessarily required for being able to evaluate digital systems, independently of what the criteria for a normative assessment may be.

1. Two Pathways Towards Trustworthy Systems and Responsible Digital Governance

In line with these considerations, we can distinguish two approaches to evaluate digital systems which are, however, in no way mutually exclusive, as we will see. Most approaches start by analyzing values and evaluating systems against them; meanwhile, our approach focuses on the question of what capacities are required to evaluate systems in the first place. We aim to briefly outline the difference and reflect on their mutual relationship.

1.1. The Value-Approach

Starting from certain values seems a promising pathway, since we seem to agree on common values. Whatever we may expect from well-functioning digital tools, we at least demand some basic normative features. An example: What would we think about a digital tool that systematically refuses access for people of a certain gender or skin color? Of course, we do not want systems to have a fundamental sexist or racist bias. Likewise, if a mortgage-decision tool always ruled against applicants named 'Peter', we would doubt its trustworthiness – no matter how accurately it might function otherwise. This line of investigation assumes that defining criteria for responsible digital governance involves a double task: First, identifying the values that matter; second, determining how to ensure that digital governance aligns with them. This perspective is, for example, central to discussions of 'Value-Sensitive Design' (Friedman et al. 2013; Hillerbrand 2021; Van de Poel 2018). Many approaches follow similar ideas though without explicitly adopting this framework.

1.2. The Capacity-Approach

This volume follows an alternative line of investigation, which we call the capacity approach. This approach begins with the assumption that, prior to any evaluation, the user must be in a position to relate to the digital tools in an autonomous and responsible way. This entails that the ubiquitous presence of digital tools must not undermine the capacity of human beings to act responsibly and to relate to each other in a responsible manner. Such responsibility presupposes that the user has the capacity to understand digital systems and their impacts at least sufficiently to position themselves in relation to them – that is, to engage with them both epistemically and practically. It also requires that the digital systems be shaped so as to be sensitive to the capacity of the user to exercise independent judgment in dealing with them.

A first critical reaction could be the following: It is not up to the digital tool whether the user is competent to use it; rather, whether the user has the relevant capacities depends on the user. This comment is entirely correct. Different people will have different capacities; they will be capable of competently dealing with such tools to varying degrees. But the problem of judging a digital system as trustworthy cannot be remedied even if everybody had a degree in computer science: First, even in a digitalized world we still need physicians, bakers, bus drivers, and philosophers. Second, quite often the most technically competent people, e.g., developers or computer scientists, don't possess the capacities to account for all aspects of a system's behavior, especially in the realm of Artificial Intelligence. The relevant capacities are, however, not only a matter of individual training and education. Rather, we can observe that certain systems are designed based on fundamental assumptions about the capacities of the intended user.

Let us explain this: If the city council sends letters to all citizens to inform them about an upcoming election, it assumes that all citizens are capable of reading. It furthermore assumes that people are capable of informing themselves about the programs of political parties and choosing which party they want to vote for. Finally, it assumes that people are capable of going to the election office at the right moment (thus, people being capable of determining the right day and time, finding the right place, etc.). All of these are implicit but necessary assumptions about the capacities of the addressee of those letters which are embedded in the very practice of sending out those letters. Thus, these assumptions are necessary conditions without which the practice of sending out those letters would not even be intelligible. These 'necessarily presupposed capacities' concerning knowledge and practices that are implicitly presupposed in a certain context, can be distinguished from capacities and knowledge about the function of the election in general. Voters are not supposed to know the details of the organization of the election in general, the mechanics of vote counting, or the legal regulations governing the procedure. The electoral authorities need to have much more knowledge about the entire process, but the voter needs some basic capacities to participate meaningfully in this practice. Similarly, the driver of a car needs some basic practical and judgmental capacities to be able to drive responsibly, while the mechanic needs more detailed technical skills that the driver of the car does not need to have to be a responsible driver. Moreover, cars should be designed in a way that the usual driving skills are sufficient to deal responsibly with a car.

Along these lines, we can observe that the way a digital tool is designed reveals which capacities are implicitly presupposed regarding its user. Some digital tools are created in a way that they require the capacities of a computer scientist to be used in a competent way. Some tools function in a way that forces the user to blindly 'trust' them, without being able to take a competent stance towards the information received from them. If a physician uses certain digital tools, these can be designed in a way that the competence of the physician as a responsible decision-maker remains untouched. But they can also be designed so that the tool effectively takes over the medical decision. If this were the case, the physician could hardly be held accountable for advice given based on those tools. Accordingly, the patient will have reasons not to trust a physician whom they suspect of receiving advice on grounds that are not intelligible to them.

The question investigated in this volume is not so much whether we can trust certain digital tools, but rather whether it will be possible to have trust in each other and in relevant institutions in a world mediated and shaped by digital tools. Whether the development of trust is possible and justified depends on whether those tools are governed in a responsible way. Whether the digital governance is realized responsibly will depend on the way in which the design of the digital tools enables human beings to deal with them, without undermining their capacity to relate responsibly to each other. This, however, depends primarily on how the design of the tools relates to the capacities of the assumed user.

In that sense, the capacity approach and the value approach are not independent of each other. The assessment of whether or not digital systems are designed in a responsible way, in the outlined sense, is a necessary prerequisite for them being morally acceptable. In some sense, this is a normative assessment that comes prior to further normative considerations. That means, independent of what users may expect from those digital systems, this would be a kind of basic requirement, and it likewise represents a basic normative quality of digital systems. In that sense, the capacity approach is not only relevant to a particular liberal framework but has much broader relevance. Of course, one can admit that the capacity approach is particular convincing for normative theories that assume the self-determination and autonomy of agents to be of central value, or that are grounded in human dignity and human rights. These concepts are, however, the starting points of most constitutions and the international human rights-framework. In that sense, such assumptions are broadly shared.

With this short outline the context in which this book is situated has been sketched. It will not result in a checklist for determining the responsibility and trustworthiness of digital governance. Rather, the book will investigate the field in such a way as to clarify the pathways through which these questions can be answered.

2. Outline of the Book

Our edition is divided into three sections.

The first section **Capacities That Enable Digital Governance** outlines the program of the approach. The contribution by **Andreas Kaminski**, **Marcus Düwell** and **Philipp Richter**, *The Capacity-Oriented Approach*, presents the underlying conceptual structure. The text builds on an idea by Christoph Hubig, namely that the ethics of technology is not simply the application of ethics to a specific domain, but rather an effort to secure the capacities for ethical reflection within the technological domain itself. This perspective reveals how prudential and deontological considerations complementarily build upon this foundational premise of all moral judgments and ethical reflection. The contribution by **Sebastian Bücker** and **Nico Formánek** *Precautions for Medical Decision Support by LLMs* analyzes on which theoretical foundations (medical) propositions made by LLMs are built, showing in which sense these systems cannot partake in the practice of "giving and asking for reasons" (Brandom). This motivates specific precautions for whenever LLMs shall be incorporated as a support for medical decision-making. In their text *Individual and Organizational Capabilities for Assessing the "Trustworthiness" of AI Systems in Healthcare Settings. The Crucial Role of Structural Empowerment*, **Oliver Behn**, **Marc Jungtäubl**, **Michael Leyer,** and **Mascha Will-Zocholl** explore the impact of AI systems in organizations. The text identifies various dimensions that are characteristic features of a structural empowerment of employees that should be guaranteed by the use of AI systems. On this basis they ask what governance structures organizations should establish to enable employees to assess the AI's trustworthiness and to ensure the continuous development and maintenance of AI-related capacities.

The second section **Governance That Enables Understanding** examines the possibilities of epistemically grasping digital systems from a normative perspective. In their article *A Right to Explanations of AI Decisions*, **Elena Dubovitskaya** and **Gregor Bosold** examine whether the right to an explanation – recognized in European frameworks such as the General Data Protection Regulation (GDPR) – may be grounded in more fundamental rights. They also explore the potential role that local explanations might play in empowering individuals affected by automated decisions, such as those used in credit scoring. **Florian Möslein** and **Michael Birkner** in their article *Cross-Chain Governance* explore the challenges in governing cross-chain blockchains, which enable interoperability between otherwise separated and heterogeneous blockchain ecosystems, thereby introducing novel complexities for blockchain governance. Governing such complexities in a responsible and trustworthy manner should, they argue, require not just technical solutions, but also address legal coordination, participant's evaluative capacities, and the design of systems capable of fostering trust. In *Blockchain-Based Governance of Financial Markets. Examining The SEC's Approach to Building Trust in Centralized and Decentralized Crypto Exchanges* **Sebastian Omlor** and **Hans Wilke** investigate how subsuming new digital systems under old legal paradigms can lead to incompatible, undesirable, and unfeasible obligations. They therefore characterize both traditional markets for equity securities and contrast these with recent markets for crypto assets. Building on this analysis, they show that the paradigm of recent markets has shifted to a point that they cannot be governed by regulatory frameworks designed for traditional markets for equity securities, thus finally outlining the necessary conditions for designing a framework that would be compatible with the paradigms of markets for crypto exchanges.

The third section **Enabling Conditions for Trust and Responsibility** explores the relation between capacities, trust, and responsibility. **Sebastian Bartsch**, **Marcus Düwell**, **Jan-Hendrik Schmidt,** and **Alexander Benlian** investigate in *Ethics and Regulation of Al Systems in Medicine. The Example of Cancer Detection* how accountability of healthcare professionals is possible if AI systems form an integrative part of medical practice. The text explores the ethical and regulatory implication of employing AI systems in this context and proposes a model for distributing accountability between the parties involved. **Andreas Kaminski** begins in *Trust in AI* from the compelling arguments that it is a category mistake to speak of trust in technology. However, as he shows, reducing trust to a narrow epistemic concept such as reliability is not the only option. Instead, he develops a conception that opens a path to speaking meaningfully of trust in technology without personalizing technology or deflating trust into an epistemic construct. In *Explainable AI as a Component of Building Trust. The Case of Regulating Credit Scoring*, **Katja Langenbucher** traces how the concepts of trust, trustworthiness, and transparency entered EU policy discourse and legislation. She then examines their relationship to explainable AI and explains how differing approaches to explanation have added a second layer of complexity. Her main contribution, however, lies in showing how various forms of explanation can enable normative responses to model-based decisions, illustrated through the example of credit scoring. In their contribution *"Trust" and "Trustworthiness"* in the AI Act, **Lucia Franke** and **Benjamin Müller** examine how the notions of trust and trustworthiness are employed in the AI Act and the EU's Ethical Guidelines. By reconstructing the meaning of trust from these documents, they conclude that it is primarily understood as reliability. They argue that this reduced understanding of trust neglects an essential dimension of trust relations – namely, freedom.

References

Friedman, Batya et al. (2013): "Value Sensitive Design and Information Systems", in: Neelke Doorn et al. (eds.): Early Engagement and New Technologies. Opening up the Laboratory. Dordrecht: Springer, pp. 55–95.

Hillerbrand, Rafela (2021): "Value Sensitive Design", in: Armin Grunwald and Rafaela Hillerbrand (eds.): Handbuch Technikethik. Stuttgart: Metzler, pp. 466–471.

Van de Poel, Ibo (2018): Design for Value Change. Ethics and Information Technology 23, pp. 27–31.

I. Capacities That Enable Digital Governance

Trust and Responsibility

Digital Systems from a Capacity-Based Perspective

Andreas Kaminski, Marcus Düwell, Philipp Richter

Abstract *Decisions in companies, public administrations, and political institutions are increasingly influenced by complex computational models. In response, numerous ethical guidelines and standardization approaches – often referred to as "AI ethics" – have been developed. Many of these approaches focus primarily on the properties of the models (such as fairness, reliability, transparency, or privacy). The prevailing assumption is that only if these properties are met can the use of such systems be considered responsible and trust in them justified. However, what is often overlooked is that this approach implies an important precondition: individuals and organizations must have the capacity to assess these systems in terms of their fairness or reliability. Given the increasing complexity – and thus opacity – of many models, it is questionable whether this precondition is actually fulfilled. This question is therefore of fundamental importance for legal and moral discourses on AI. In our article, we introduce the idea of a capacity-oriented approach. We show how virtue-ethical considerations and moral principles aimed at agency and autonomy can provide the foundation for this approach. Building on these reflections, we propose a revised requirement for digital governance and raise the question of how governance can be designed to make systems responsible in such a way that their use can be considered trustworthy.*

1. The Usual Approach and its Critical Prerequisite

In many discussions surrounding digitalization and artificial intelligence, we encounter demands for *trustworthy* systems that have been designed *responsibly* (Pekka et al. 2018; Floridi 2019; Thiebes et al. 2021; AI Act 2024). Attention here primarily focuses on the normatively relevant properties of such systems, such as being fair, reliable, or health-promoting. According to this view, a system is considered trustworthy if it possesses certain properties, for instance, being just and reliable (Bisconti et al. 2024; AIEI Group 2020). If it exhibits these properties, it can also claim to have been responsibly designed, and its deployment may potentially be regarded as justifiable. However, this line of thought explicitly or implicitly presupposes that individuals, communities, or organizations are able to relate the functionalities and performances of such systems to normative considerations relevant to them (e.g., values, beliefs, principles, or rights).

Thesis 1: Reasonable trust in systems and the acceptance of responsibility for these systems require the capacity to evaluate whether such systems are worthy of trust and can be used responsibly.

A person, community, or organization that wants to evaluate whether a system is just or reliable, promotes health or autonomy, must understand (1) *how that system works*, at least in principle, and must also understand (2) *how the system relates to normative aspects*. The twofold prerequisite is essential for the evaluation of technical systems in general, and digital systems in particular.

This raises two questions. The first is a *factual* question: Is this prerequisite fulfilled with regard to a specific technical system? And if so, for which actors and to which degree? This question can, for example, be examined within the context of sociological studies. The second is a *normative* question: What conditions should be in place to enable individuals or organizations to evaluate technical systems? How can we ensure that this critical condition is fulfilled? How should systems be designed to support individuals or organizations to achieve this capacity? And furthermore, how can organizational environments be created that foster the development of such capacities and prevent technical developments that hinder them? Both questions lead directly to the issue of digital governance on the one hand and an enabling-capacities approach on the other hand.

Consider a system for granting loans or a medical system that diagnoses illnesses and recommends therapies. We assume here that both systems operate based on machine learning methods. For individuals or organizations to be able to evaluate loan decisions or medical diagnoses, they must first understand which factors were decisive for the loan approval or the diagnosis. However, this alone is insufficient; they must also be able to assess the quality and appropriateness of the decisions. Imagine that a person's loan application has been rejected, or that radiation therapy is recommended due to a tumor diagnosis. To evaluate such decisions or recommendations, individuals must understand whether they are well-founded, fair, or beneficial for health. This brings us to the role of the capacities for understanding and evaluating. Beyond this, the question arises whether the organizations operating these systems enable reasonable interactions with them. Can objections, for example, be raised and addressed appropriately if a loan is rejected without appropriate reasons? Do the involved medical experts have sufficient time and knowledge to engage with the systems and its decision, to question it, learn from it, and, if necessary, to reject its recommendation? Such questions pertain directly to the digital governance of organizations.

2. The Enabling-Capacities Approach

Here, this book proposes a shift in perspective: Until now, ethical and legal debates – especially in the context of AI – have primarily focused on directly evaluating the properties of technical systems; judging whether these systems meet certain preferred conditions. However, we argue that to fully grasp what it means to regard certain system properties as good, better or worse, we must first take a step back. Before we may focus on particular properties, we need to reflect on the task of evaluation itself.

Our claim is this: Whenever a system is evaluated, there is always someone who is making that evaluation through a value judgment. Evaluative judgments, when made on a reasonable basis, presuppose that certain conditions are met. The subject who makes the judgment must possess specific capacities and be able to exercise them appropriately. This raises the question of what those capacities are, how they can be cultivated, and what conditions are particularly conducive to their development and exercise. Evaluating technical systems therefore requires *capacities* that are a prerequisite for evaluation; and it raises the question of the conditions that promote, facilitate, impede, or prevent the development and exercise of these capacities.

Figure 1: From a focus on properties to a capacity-based approach.

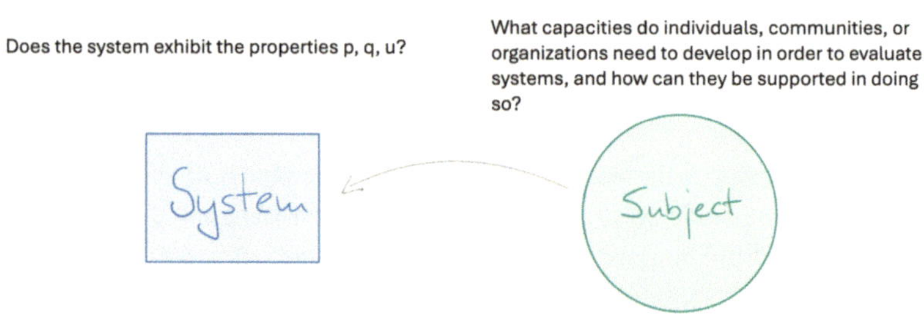

In principle, the capacity-based approach has always been assumed within the context of examining the properties of a technical system. What we are doing here, first, is to make this seemingly unproblematic premise explicit, since it is by no means always uncritically fulfilled. Second, we are no longer directing our normative questions solely at these properties, but primarily at the conditions under which capacities are developed to identify and evaluate those properties. These conditions, after all, can themselves be shaped, politically demanded, and technically supported. Third, this may lead to considerations that urge to come up with a more integrated normative view on the normative assessment of those technical systems.

> **Thesis 2:** Not only the properties of technical systems can be subject to ethical, political, and legal evaluation, but also the conditions under which such evaluations can take place. What is evaluated, then, are the conditions under which capacities are developed to identify and assess these properties. After all, these conditions can themselves be shaped, politically demanded, and technically supported.

Although, as mentioned before, this shift in perspective is not without precedent. One of our most important sources of inspiration lies in philosophical ethics of prudence dating back to Aristotle in ancient times. Within this tradition, as we can say, attention is paid to the capacities required for making appropriate value judgments. At the same time, consideration is given to the conditions conducive to prudent and ethical action.

Some words about *ethics* may be necessary here. Ethics is here not understood as a set of customs or rules of behavior but as a philosophical discipline – which is a common view since ancient times. As a philosophical enterprise, ethics forms a discipline that in a systematic way reflects on questions like 'What shall I or shall we do?' or 'How shall I lead my life?'. In that sense, morality, moral norms, duties, rules, or principles are the topics on which ethics reflects. Philosophical ethics systematically aims to understand what is so specific about moral norms and values, how we can speak about them, and whether those moral requirements are only expressions of subjective views – only opinions – or whether moral requirements of right actions can be justified in intersubjective context. But moral norms and principles are not isolated in human praxis. We are not only confronted with moral questions but in normal life we ask as well, how to live one's lives? What is important for us in life? What makes us happy? These are questions where some variety of answers are possible. While some would be ultimately happy to play several hours of football a day, for others that would be a nightmare. Nevertheless, there seem to be some considerations that seem to be quite generally valid. Being never able to fulfill their dreams and aspirations is something that makes people unhappy, independent of what the dreams and aspirations may be. Permanently ignoring one's talents and not developing them seems not a good idea for human beings in general – even so it may not be immoral to do so. Thus: There is a dimension of human life that does not coincide with morality but about which ethics has something to say.

This field is often called an 'ethics of prudence'. Today, prudence or prudential reasons are often seen in direct opposition to morality, since prudence is understood to deal with what one could ideally achieve for oneself. This is because prudential reasons seem to be derived from mere instrumental means-ends-calculations. If we say a person behaves prudently, we seem to articulate the suspicion that the person aims at achieving whatever selfish ends they might have – and the smart guy is not necessarily the good guy. On this view, ethical theory must establish justifiable constraints to counter the prudent pursuit of self-interest (Kaspar 2011: 311 et seq.). However, this is only half of the story, prudence was and is seen also as a complex intellectual and practical virtue, being bound to certain moral values or complementing them practically (Annas 1995: 245, 251; Luckner 2005). There is a philosophical tradition of diverse thinkers such as Aristotle, Machiavelli, or Descartes that can be called the "ethics of prudence" (Luckner 2011; Benner 2010; Cimakasky and Polansky 2012). This tradition is especially helpful for tackling the challenging blend of descriptive and normative questions in applied ethics (Richter 2018: 47–51). This ethical approach does not necessarily conflict with principle-based ethics or universal moral norms, rather it complements them by considering the processes and criteria through which individual subjects deliberate, judge, reason, and act according to ethical reasoning (Hubig 2007: 140; Fremstedal 2018). We will come back to this relationship between ethics of prudence and morality below.

2.1. Applied Ethics as an Enabler of Ethical Reasoning (Hubig)

German philosopher Christoph Hubig developed an approach to applied ethics which heavily draws from ethics of prudence (Hubig 2007: ch. 5; Hubig and Richter 2015). By descriptively acknowledging moral pluralism, both, in practical reality and in moral theories, Hubig draws our attention to the common ground of all ethical theory: All ethical approaches are focusing on different aspects of free human action. Here, freedom is not understood as human beings acting without any constraints, like external restrictions and forces which they do not have chosen, or internal limitations, like inhibitions, fears, or lack of motivation. However, we must assume that human beings can reflectively relate to their behavior and ask questions like 'What can I do? What do I want to do? What am I obliged to do?' Freedom in that sense is necessarily presupposed since ethics deals with qualifying our action as prudent, right, or good. All ethical approaches must presume that there is free action since otherwise all ethical reasoning would be futile and practically irrelevant (Hubig 1995: 113). This holds for moral theories about universal duties and for prudential approaches.

Ethics of prudence starts, as we can say, with one general assumption about practical freedom: It is not within our power to do everything as we please, but neither is everything beyond our control. Practical freedom is understood as a reflective behavior of setting oneself in relation to identified determinants, thereby gaining clarity on the available options for action (this also could be called 'to orientate oneself' in the sense of getting oneself ready to act, see Luckner 2005: 9–30). Here, freedom is both an ability and its procedural actualization in deliberation, which involves learning about what one's abilities and options are, and thereby getting oneself ready to choose what is normatively good in the long run. It is an epistemic subtask of prudent judging and acting to take precautionary measures against real or potentially upcoming dependencies (a lack of practical freedom) and to extrapolate these from empirical knowledge as well as to actively obtain the necessary knowledge about immediate and future circumstances of action.

However, prudential ethics does not simply assume freedom as a matter of fact. Freedom is vulnerable, can be restricted, and is endangered. Therefore, it asks how freedom can be preserved or enhanced in practice. A central aim of all prudent thinking and a general good to agents is therefore the long-term preservation and increase of the ability to act because, for instance, according to Aristotle, human activity in the sense of the associated virtues must be considered intrinsically valuable from a practical perspective. Without going into detail about a foundational argument, it is quite plausible to take the 'ability to act' as a general good of all prudent deliberations: in general, it is true that situations need to be avoided in which one is forced to react without free deliberation, having only little or no other choices, since it is bad to be unable to do what one really wants or what would seem reasonable. We can say true prudent deliberations are only those that reach their specific ends and at the same time allow for further prudent deliberations in the future. Prudence could be seen as a capacity-saver and enabler. As far as possible, a prudent person would avoid severe restrictions or the loss of the ability to act. Known or deducible examples of the severe limitation or loss of the ability to act are for instance: severe psychological or physical dependencies (such as addictions), but also lack of income, as well as socially or naturally induced disasters (such as wars or flooding), technical constraints (*Sachzwänge*) as well as lack or failing of infrastructure. All of those should be prudently avoided by taking precautionary measures even if this is often only realizable via collective efforts.

Let us imagine, for example, that it is a legacy system that has been maintained by different people over long periods of time and is poorly documented. Major changes to such a system may either appear to be unfeasible, with the result that the only option is to try to keep operations running. Or, to give another example: a system has become so dominant that there are no real alternatives on the market. Cases such as these are to be avoided by acting prudently, as they severely limit or even exclude the possibility of prudent behavior in the future.

Now, to avoid all these anti-prudent restrictions could be seen as normatively demanded by rational egoism being in favor of an individual's own agency to the possible detriment of others. However, this is not necessarily the case. We can consider the ability to act as a good while we can see prudence as a way of reasoning to maintain this good. If we see it like that, prudence generates the preconditions for possible moral action in the long term, since without the ability to act there is no ethical reasoning and no moral action according to it. So, to achieve moral action according to universalist morality there must be prudent thinking (Hubig 2007: 129, 131 et seq.; Kaspar 2011: 320 et seq.). We find that line of thought, for instance, also in Kant's universalist moral theory when he speaks of an 'indirect' duty to "secure one's own happiness" (AA IV: 399), since a lack of practical possibilities due to crisis, poverty, fear, or dependency and so on "might become a great temptation to transgression of duty" (AA IV: 399). So, moral reasoning, even according to principle-based ethics like Kant's, involves provision and precautionary measures to sustain the ability of moral action. Here, we may suspend the question of whether true prudence is already bound to moral virtue, as for instance Aristotle took it (in contrast to Hobbes and Kant). For us, it is sufficient to assume that prudent thinking is bound to the end of sustaining agency in persons, groups, or states, which is a practical precondition to deliberate about actions also in an ethical way. The point is, first, there must be at least some options to form decisions and actions according to them, then, in a second logical step, these can be evaluated from an ethical point of view. Without options, no actions. And without actions, no object of ethical reasoning. In this respect, actions (or system properties) may be judged as morally good or bad. However, to form these judgements we also have to admit that it is good that the world contains at least some actions (or persons judging system properties); otherwise, ethical reasoning would be futile and meaningless. For sure, prudent action is not the whole of ethics, since it does not offer founding arguments for principles holding universal moral norms. However, universal moral norms or goods can only be fulfilled if their addresses are able to act now and in the future.

To come back to Christoph Hubig's approach to applied ethics, the common ground of all the diverse ethical reasoning according to a standpoint of ethics of prudence is to sustain the ability to deliberate and to act freely. If this is combined with foundational theories of universal morality, e.g., Kantian ethics of autonomy, we find two complementing aspects of free agency in agents being able to ethically reflect under conditions of system induced chances, dependencies, and restrictions (Hubig 2007; Hubig and Richter 2015).

2.2. Option and Legacy Values

Since freedom exists in relating oneself to not or only hardly changeable determinates, and in choosing in the remaining leeway of options considering some of them as better or worse (cf. Raz 1988: ch. 14), all agents need a variety of options to choose from, otherwise there exists no practical freedom. We can call this a view of practical freedom "from outside". However, considered "from inside", agents also need to be able to consider themselves a constant self over time and during different sequences of action, for if my standards of evaluation changed from one moment to the next, no action extending beyond the immediate moment would be possible. This capacity – being the author of one's actions – depends on social preconditions (ibid.). For instance, if a society allows violence or disrespect, it is nearly impossible to develop long-term plans or to inculcate agency supporting virtues like practical self-efficiency, self-respect, or trust in one's own abilities.

The preservation of the ability to act, regarded as the highest aim of prudential ethics in this sense, according to Hubig, presupposes two fundamental types of values:

1) "option values": the preservation of options for action;
2) "legacy values": the preservation of being a subject, namely the ability of cultivating and maintaining a personal identity that informs a way of life from past to present (Hubig 2007: 141–145).

Shaping technology requires options. Without at least two options (a or b), there is no space for the design of technology. When no options are given, for instance, because practical constraints exclude alternatives, then evaluations are not feasible or meaningless. One can still arrive at the conclusion that a certain technology is 'bad', but this judgment may remain inconsequential.

Let us imagine, for example, a software that has become the standard, and which one considers poor in numerous respects (it is too expensive, does not adequately protect data, and restricts possibilities for action). The company that produces the software is aware of this criticism; however, it does not respond, or responds only selectively to the deficiencies, because it knows its status as an (industry) standard. The same applies to models for evaluating creditworthiness that have become quasi-standards, as well as to development paths (such as in electricity or mobility) that can no longer be easily changed.

Furthermore, the assessment of technologies requires that subjects can relate to technologies. This means that subjects must know that and how they are affected by these systems; for this, they must be aware of the effects of the system. However, this is not the case if the effects remain below the threshold of perception or cannot be understood. Furthermore, subjects must not be influenced by the systems in such a way that their standards become incoherent.

It is important to see that in this perspective the mere existence (and maintenance) of options to choose from and the preconditions for generating subjects (e.g., family structure, social services, education at schools or universities) is prudently good and indirectly morally required even if these possibilities will not actually be used with specific benefit in the near future. The mandatory provision to keep these possibilities in stock does not necessarily involve concrete action plans of how to make use of them; this should be left to the individuals of whom ethical reflection in one or the other way is expected.

Let us imagine a system that assesses the creditworthiness of individuals without their knowledge that such a system is being used for this purpose. They also may be unaware of what data is taken into account and how it factors into credit decisions. Such a system, for instance, could monitor the browsing history of individuals, the history of their addresses, and much more without them being aware of this surveillance or without them understanding its impact on the credit decisions being made. This is why legacy values are required in order to evaluate technological systems.

Option and legacy values are meta-values (not specific rules or principles) functioning as what in ancient Rhetorics is called "topoi". Topoi are criteria which need to be considered always when dealing with certain topics or problems. However, they always need specification based on empirical knowledge about the real-life situation now dealt with (Hubig 1990: 134, 140 et seq.; Richter 2017: 190, 195 et seq.). If we follow Hubig's approach, we may accept the pluralism of moral approaches and diverse ways of ethical reasoning (which is one of the preconditions of free agency in a modern society) but follow a perspective for applied ethics which is valid universally: A technical system, may it be in health care, internet-based technologies, or traffic infrastructure, could only be morally acceptable or good if users are and kept able to relate themselves to the systems infrastructure, to gain practical freedom, and the possibility for ethical reflection. According to option values, technical systems must grant the users a scope of possibilities for creative redesign, rededication, critique, discussion, or rejection of intended ways of use. If there is little opportunity for choice in the sense mentioned before, e.g., if the determinants or system functions cannot be identified and no alternate ways of use are visible, then there is little chance for users to reflect on their behavior and that of the system in a broad and ethical way.

However, how to achieve this in a technical system is not straight forward like an easy deduction, since there are conflicting goals, since technical systems relieve and facilitate procedures which otherwise needed to be done manually by human action. Therefore, a prudently wise balance between giving up freedom by making use of system infrastructure and maintaining freedom while relying on a system is required (Hubig 2007: ch. 6; Luckner 2005: 164 et seq.). This also holds for further specifications of the prudential approach. Topoi of action enabling capacities need to be developed, both in a theoretically and empirically informed way. In a next step these considerations need to be specified in detailed studies about the technical systems in question. If we consider algorithmic based digital systems, then the topoi understandability and controllability need to be considered since these capacities function as preconditions of users' practical freedom under the restrictions of these types of systems. For without the ability to understand technical systems, it is impossible to recognize how they shape and transform practices and decision-making contexts; without the controllability of technical systems, the systems cannot be shaped or steered.

2.3. Prudence and Morality (Aristotle and Kant)

We now have presented some reasons why the freedom of agents to relate to technologies in a free and controlled way can be justified from a broadly speaking Aristotelian approach. One may wonder whether these considerations were only plausible if one subscribes to an Aristotelian concept and whether this is not a type of ethics that is at odds with a modern concept of morality which is more focused on universal moral principles. This is particularly relevant since central modern concepts like human rights and human dignity which form the cornerstone of a liberal worldview are based on those universal aspirations. Philosophically, the most obvious comparison to discuss these questions is the comparison between Aristotle and Kant. In contrast to prudential reasoning, a principle-based approach is often understood to be focused solely on the justification of universal moral principles, thereby neglecting the contingencies of practical reality in which these have to be specified and employed wisely. This applies especially to how Kant is commonly understood. Thus, the difference between context-sensitive prudence and strict universal moral principles seems to be the difference between Aristotle and Kant.

We propose a different take on the relation between Aristotle's idea of prudential reasoning and Kant's moral philosophy. Already in the *Groundwork*, Kant emphasizes that agents have to see themselves under three imperatives, we could call them (in modern terminology): instrumental, eudaimonistic and moral imperatives. Kant-scholarship focused primarily on the moral imperative, which is the only one categorically valid, that is the only imperative that is binding for agents under all circumstances. But Kant stresses that there are three imperatives (see for this interpretation of Kant: Steigleder 2002: 23–58). Whenever an agent is committed to realize goals, he must also be committed to the means which are necessary to reach these goals. If it is the goal of an agent to become a successful guitar player and if she is really committed to this goal, she has to practice regularly, otherwise she cannot claim to be committed to this goal. This is imperative for the agent in a strict sense, an agent that would assume that he is committed to the goal but is not as well committed to the necessary means would not understand himself consistently. This imperative is only in that sense not categorically valid, since it is not necessary for the agent to hold this commitment to this goal, this commitment is contingent; the agent can just decide that she does not want to become a guitar player any longer. But as long as the agent is committed to the goal, he is necessarily committed to the necessary means. Since there can be more than one means appropriate to reach the goal, the agent must also be committed to be able to reflect on the appropriate means and the relevant capacities to realize those reflections. The agent must furthermore be committed to strive for happiness. If the agent were not under the eudaimonistic imperative, he would not have any reasons to prefer one goal to the other. If the agent did not have any reasons for choosing a certain goal, it would not be plausible why he could be harmed if anyone hindered him to realize his goal since the choice of this goal would be totally random. The ability to form maxims must be guided by this commitment to eudaimonism, otherwise it is not evident how maxims ever could be generated. Thus: the agent is under the eudaimonistic imperative, but this commitment as well is not unconditionally binding but is restrained by the moral imperative that alone is unconditionally valid. There may be conditions where the agent is confronted with duties that are really morally important even so, it will come with restrictions on the happiness of the agent – sometimes the moral demands can really come with far-reaching negative consequences.

Even if in that sense morality can be opposed to the happiness of the agent, the categorical imperative must be embedded in the two other imperatives, otherwise various aspects would not be understandable, here only two should be mentioned: First, Kant stresses that the moral imperative obliges the agent to investigate whether maxims are acceptable for all other agents as well. This already assumes the other has an order of means and ends that is meaningful for the other. Second, Kant stresses that we have a duty to promote the happiness of others. This also assumes that happiness is meaningful for others as well. For Kant, controlling the circumstances of my action and to be able to form an informed judgment is of central importance. The moral imperative is understood as an articulation of our autonomy as rational beings. We should see us as morally bound because the categorical imperative is justified by our consistent self-understanding. This emphasis on the 'capacity to judge' is so to speak the cornerstone of his entire philosophy (Longuenesse 2001). For our context, this is important since the Kantian approach does not only entail that the agent is only capable of deducing concrete norms form general principles. It rather also entails a broad range of capacities that enable the agent to orient oneself in the world. Being able to place oneself in the position of everybody else – as the application of the categorical imperative entails – presuppose that the agent has the imagination to transcend the own position. Being able to form judgments about happiness and morality entails the capacity to imagine what course of actions are possible. In the *Critique of the Power of Judgment* Kant emphasizes the importance of aesthetic judgments for orienting oneself in the world. One could add: Only if the agent can imagine new ways of how she can act, the choice between different courses of action is meaningful (Düwell 1999). In that sense, the focus on a concept of morality as categorically binding norms and principles is dependent on the preservation and cultivation of the power of judgment.

2.4. From Immediate Situations to Long-Term Effects

These considerations are particularly significant in the domain of technology. The necessity of integration (of prudential reasoning and principle-based morality) becomes even more obvious if we think about more complex forms of moral, political, and legal considerations, where we are not only faced with questions about moral duties in single situations, but also where structural questions, long-term effects, and uncertainty about the consequences of actions have to be considered. In all those contexts, the agent first has to try to understand what the current possibilities of actions are, which consequences they may have and how they will affect other agents. All of that presupposes that the capacity to judge is developed and that the agent is in the position to exercise this capacity.

In the line of these considerations, one could emphasize that theories of (human, individual, subjective) rights should also be seen as embedded in such a broader concept of enabling capacities. It has already been mentioned that those rights are the cornerstone of modern societies. The link between rights and enabling-capacities becomes plausible if we consider that rights do not come in isolation. Only in exceptional situations, the protection of only one right is at stake (questions of life and death, torture, emergency situations, etc.). It is, however, much more common that we are faced with competing rights claims or questions where we have to decide about long-term, cumulative, or indirect effects of actions on rights. Quite often there are tensions between the prima facie rights of people to exercise some liberty and rights of others that can only be realized if some liberties are restricted. In all those conflicts questions about hierarchy, urgency and priorities between rights claims are on the table. This leads to questions of consistency in the interpretation within the systematic connection between rights (see Gewirth 1978; Düwell, Graf Keyserlingk and Richter 2025). At least in three respects questions of enabling capacities are relevant. First, commitment to rights implies a commitment to the necessary conditions for the realization of such a right. If we (individuals, groups, states) are committed to a general right to freedom of movement, we must as well be committed to protect and support the conditions that are required to exercise this right. Second, this now implies not only conditions that are necessary for an individual to realize his individual rights but also the collective conditions under which it is in general possible to exercise those rights. In case of freedom of movement this implies at least some form of infrastructure that is required to exercise this right. Third, regarding various rights, it is neither *eo ipso* evident what the necessary conditions for a successful exercise of these rights are nor how the importance of this right can and should be weighed against another right. For that reason, it would at least be required that there are procedures and decision rules which are required to form an informed and collectively accepted decision. All the three dimension show that there are enabling capacities required under the assumption that one has some commitment to certain rights.

3. The Role of 'Understanding' and 'Explaining'

The evaluation of systems, whether in specific situations or from a long-term perspective, is challenging due to their partial opacity. This opacity first pertains to their internal functioning, which we can refer to as *model opacity* (Humphreys 2009; Beisbart 2021; Burrell 2016).[1] However, there is a second type of opacity that has not yet been explicitly addressed or conceptualized: Opacity can also extend further to include the consequences and side-effects of these systems, as well as the quality of those outcomes. This can be termed *pragmatic opacity*. This opacity does not concern the model itself, but rather the effects of the model in (real-world) application contexts.

A simple example would be a web service (e.g., web search) that adapts the selection of information to the users (modeling user preferences). Model opacity would occur if, for instance, we cannot anticipate whether, how, and why a (slight) change in input data would alter the model's results. Suppose a user is 35 years old, female, and lives in Berlin. The model is opaque for us if we cannot comprehend (a) whether, (b) how, and (c) why something would change if she were 33 years old or lived in Munich instead.

Pragmatic opacity, in contrast, arises if the user does not recognize (a) that and how the system adapts the information selection for her, or (b) how appropriate (or beneficial) this selection is. Clearly, it is essential not only to identify these different explanatory factors but also to weigh them, particularly regarding their significance for specific (material, fundamental, ethical) values.

In the case of model opacity, even if the mathematical function of each element (neurons, cost functions, hyperparameters, etc.) can be examined and traced, this alone does not yield an understanding of the overall behavior of the model – as evidenced by the fact that typically one cannot anticipate the behavior of the model, even with only minor adjustments (Kaminski et al. 2018). Pragmatic opacity, on the other hand, refers to situations where the effects of a system on its environment cannot be easily understood or reliably evaluated. This concerns not only the direct effects produced by the system but also the long-term social, political, and other implications associated with its use.

1 This is usually referred to as "epistemic opacity". However, this term is typically understood to refer only to the opacity of models, and not to their pragmatic opacity. For this reason, we prefer the term "model opacity" over "epistemic opacity."

Current research primarily focuses on minimizing model opacity, while neglecting the realm of pragmatic opacity. As a result, the significance of models for their respective fields of application is not adequately considered. (Pragmatic opacity means precisely that one cannot assess the quality of system outcomes; quality with respect to the fields of action deemed relevant.) The concepts of 'explaining' and 'understanding' thus appear from two distinct perspectives: firstly, in relation to the model and the previously discussed model opacity; and secondly, regarding the significance of systems in terms of how they transform our practices; that is, with respect to pragmatic opacity.

First: In terms of models, we therefore understand *explaining* as the ability to elucidate which factors actually caused a model result. By contrast, the capacity to *understand* the model requires more: This ability consists of being able to anticipate model behavior under counterfactual assumptions.

Figure 2: Explaining and understanding of opaque models.

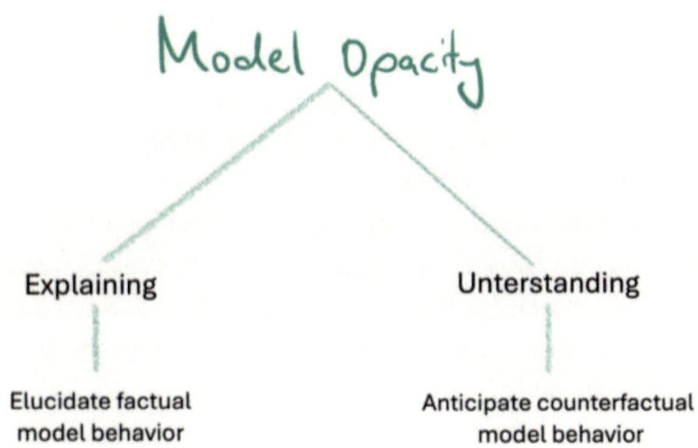

In the case of a credit scoring model, this means that decisions are explained when we are able to identify the decisive factors that led to an actual approval or rejection. We understand the model to the extent that we are able to anticipate, counterfactually, what outcomes the model would produce under altered conditions (e.g., marital status: two instead of three children; place of residence: Bonn instead of Gelsenkirchen; age: 35 instead of 50 years).

Second: With respect to our practices, the aim is to clarify how the use of systems transforms those practices. The capacity to *explain* this consists in articulating how a real system alters our actions; for instance, by reducing or expanding our options, transforming our value orientations, or introducing effects that remain unnoticed by us.[2] *Understanding* the use of such systems, analogously, refers to the capacity to anticipate, counterfactually, what changed significance a modified system would have for our practices.[3] This latter capacity, in particular, is crucial for the evaluation and design of such systems.

Figure 3: Explaining and understanding how systems affect our practice.

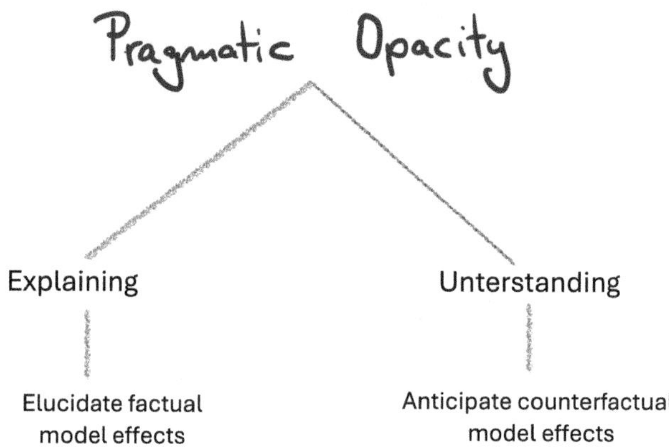

Consider again the case of a credit scoring system: Explanations in this context might consist in identifying why, since the system's introduction, individuals have begun to optimize their creditworthiness in different ways; or why fewer older individuals apply for credit, and how this shift affects economic conditions. The ability to practically understand the use of such systems would mean being able to anticipate, for instance, what consequences a different weighting of certain factors would have for small businesses, families of a given type, and the like. Or what risks might arise for precariously employed individuals if credit were granted to them more easily, and so on.

What remains open, however, is *for whom* models or systems are supposed to be explainable and understandable: individuals, groups, organizations? Or more pre-

2 To our knowledge, there are no papers that directly address pragmatic opacity; nonetheless, there are several lines of thought that point toward aspects that are relevant for this concept. These include works that discuss the use of digital systems operating below the threshold of perception (Nordmann 2008), or that examine the effects of anonymous forms of collectivization ("anonyme Vergemeinschaftung"; Wiegerling et al. 2008: 82).

3 We adopt Beisbart's suggestion here to tie understanding to the role of counterfactuals, even though Beisbart introduced this term with model opacity in mind (Beisbart 2021: 11657).

cisely, with regard to levels of expertise: laypersons, experts – and if so, which ones? More on this later.

4. When are Systems Trustworthy?

Technical systems are trustworthy when they fulfill the relevant values within a given context. At first glance, this statement may seem vague or empty, as it is formulated in abstract terms. However, it is in fact essential to relate trustworthiness to other values that are relevant to consider in specific situations.

Let us walk through this step by step:

Trustworthiness is related to (other) values. This becomes clear when we look at interpersonal relationships. When we trust a person, we expect them to be honest when it matters; to be reliable, kind, courageous, or just when the situation calls for it.

As this shows, trustworthiness is not simply one value among others. Rather, trustworthiness is directed toward the fulfillment of other values. Thus, it is a (higher-order) value of systems that refers to other (first-order) values. Which values these are depends on the situation at hand. Let us compare this to interpersonal trust. If I trust a friend to speak up in the face of an unjust and intimidating colleague, I am relying on her courage. If I trust a friend to pick me up on time so I can catch my flight, it is his reliability that matters in that moment. In such situations, the friend is trustworthy when he acts reliably or, as in the earlier case, courageously.

The same applies *mutatis mutandis* to technical systems. Here, too, trust is directed at trustworthiness, and trustworthiness refers to the relevant values and meta-values (such as reliability, fairness, or privacy, autonomy) and system functions (such as promoting health or enabling prosperity).[4] As a consequence, in order to assess a system's trustworthiness, individuals must evaluate whether the system fulfills the respective first-order values.

Three points, however, must be kept in mind:

(1) The way we talk about values might be understood in a reifying manner – then it may seem as if such language commits us to a reality of values. However, we do not intend to touch on that question here. When we speak of values in what follows, we

4 Hartmann initially describes trust in other people as something that emerges in a relationship that it is not only about trust itself. Trust is part of a practice in which it contributes to realizing values other than the value represented by trust itself for Hartmann (2011: 15–18). The same assumption is often applied to technology. In the following, however, we take a slightly different approach than Hartmann, who develops the relationship between values based on the distinction between instrumental and intrinsic ones.

are referring to the practice of evaluating. In this practice of evaluating, we adopt different standpoints. We assess a system in terms of how reliably it produces a certain outcome, how securely it protects data, or how fairly it distributes resources.

(2) The way we spoke about values may suggest that these values operate on the same level. This is by no means the case. We propose distinguishing at least four levels of standpoints, they refer to the various standpoints we can take when assessing systems.

The first level of standpoints involves evaluating systems in *material* terms based on the role or function a system is intended to fulfill, such as prosperity, health, safety, or connectedness. The second level refers to the *formal* assessment of common moral virtues such as justice, courage, or sincerity. This formal perspective often focuses on the material aspect, such as the *just* distribution of *health contributing* resources. What becomes apparent, therefore, is that there is an order among the different levels. This continues with the next aspect as well, for the third level refers to *value enabling values* among those are the option and legacy values. They ensure our capacity to act because they refer to the possibility to revise the course of action and to relate to the effects of a system (which presupposes its explainability and understandability). Thus, the third aspect is connected to the first two. The formal evaluation of systems focuses on the material goods involved such as the fair distribution of resources (e.g., credit) or the sincere and reliable handling of data. The third perspective ensures that systems can be shaped in accordance with the first two viewpoints. This order is finalized on the fourth level: *Trustworthiness* and *responsibility* constitute the *unity* of these values (cf. Kaminski 2021: 396–398).[5] They encompass the other aspects in as much as models can be considered trustworthy and their deployment accountable if and only if they promote, or at least do not violate, material, formal, and value enabling values.

5 Trustworthiness can be understood as a unity, since it integrates the values that are relevant in a given context. To call a person or a system trustworthy is to say that this trustworthiness is constituted by other values that matter in the situation at hand. A person or system is trustworthy, for example, when they are reliable, transparent, or fair in contexts where such qualities are decisive. In this sense, trustworthiness is not an isolated trait but a unifying quality that brings together other values.

Figure 4: The order of viewpoints for the assessment of systems.

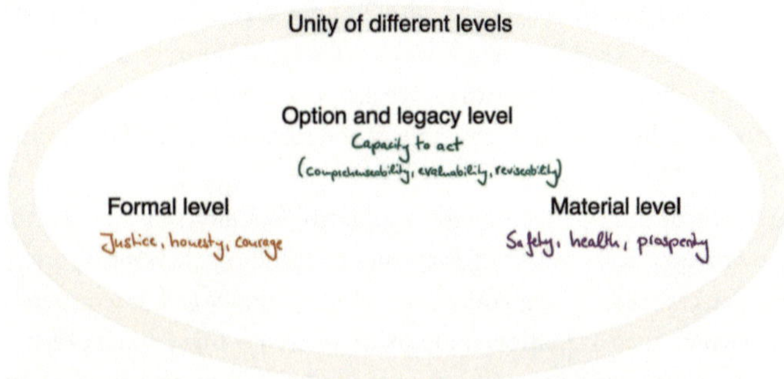

The discourse on values may suggest that it is always already clear what these values are and how we determine whether they are being violated. Instead, we adopt the distinction introduced above between models (and the corresponding model opacity) and systems as they are embedded in contexts of action (pragmatic opacity). With respect to models, it indeed seems straightforward to evaluate them by measuring them against certain values such as reliability, safety, or justice. The same, however, does not hold for the deployment of systems in practice. In these cases, it is often not clear which values are relevant; nor is there necessarily a consensus on how to interpret the values in question. Moreover, the very understanding of what counts as a value and which values we consider relevant may itself be altered through the use of these systems. In such situations, what matters is our capacity to understand how systems may transform our practices, so that we may orient ourselves accordingly.

Thesis 3: Trustworthiness refers to a different level of assessing a system. It unifies the lower levels and their viewpoints. If we come to the conclusion that a system is trustworthy, this means that we have evaluated it positively on the other levels – for example, it fulfills its material function of promoting health, and does so in an exemplary formal manner, such as being fair and reliable. In addition, we have access to this system and are able to shape it (option and legacy value).

When a person, community, or organization seeks to assess the trustworthiness of, say, a medical model, they must first evaluate whether and to what extent the model fulfills other values that are relevant within the respective domain, such as health, justice, or reliability. The degree to which these first-order values are promoted or realized by the model determines its trustworthiness. However, if we want to understand how systems transform our practices, it is no longer sufficient to measure models in this way. Two ex-

amples may illustrate this point. A system used in the care of elderly or ill individuals may operate reliably and promote health, yet by primarily treating these individuals as subjects to be relieved of burden, it may diminish their capacity to act (Wiegerling et al. 2008: 75). In doing so, it would violate a basic value (whether such a violation is acceptable or ought to be compensated is a further question). Or consider a system designed to support medical professionals in diagnosis: it may alter their competencies – undermining or developing certain skills, calling into question justified or unjustified self-confidence, and so on. In both cases, even from the perspective of the model, it may not be clear what the relevant values in our practices are, how they should be conceptualized and related to one another, or how they might shift through the use of the system. What we require, then, is an understanding that takes into account the pragmatic opacity of the contexts of action.

5. Responsibility and Digital Governance

'Responsibility' is a normatively underdetermined concept and therefore must always be specified through additional criteria. Responsibility can be conceptualized as follows:

'A is responsible for X to B by appeal to normative standpoint Z.'

There is thus an acting person or institution A who takes responsibility for or is held accountable for an action X. Furthermore, there is a person or institution B to whom the responsibility is owed or by whom it is demanded. This relationship is based on a *normative foundation* Z (e.g., a social norm, a law, etc.) that gives further definition to the responsibility. This basic schema can describe both moral and legal responsibility.[6] The concept of responsibility thus allows for the description of a responsibility relation, but the normative basis for attributing responsibility must always be specified; it does not follow from the concept of 'responsibility' alone. Therefore, appeals to 'responsibility' are always at risk of being used in a purely rhetorical way.

There have been various attempts to define the normative basis of responsibility by tying the attribution of responsibility to the realization of certain values. The approach proposed here, however, assumes that a prior question must be addressed: namely, whether agents are capable of understanding themselves as subjects of responsibility in the first place. It is thus suggested that the attribution of responsibility be expanded to include a reflexive dimension.

---

6 In one case, it is a matter of moral norms, in the other of legal norms. On this, see Werner 2002: 521et seq.

Thesis 4: For the capacity-based approach, accountability does not (only) mean that systems are measured against certain values. Rather the key question is whether it is possible to relate systems to our practices, norms, values, and goods in an evaluative manner. This, in turn, presupposes digital governance that enables and secures precisely this possibility.

The primary issue is not whether digital systems realize particular values, but whether the acting person A and the entity B – toward whom responsibility is assumed or by whom it is demanded – are capable of determining the normative basis Z. This higher-level dimension of responsibility is necessary because, in the context of digital systems, the normative foundation to which one can appeal is not self-evident. We do not yet know which digital systems are justifiably accountable, and we must first be enabled to make such determinations.[7] This capacity for judgment must be safeguarded and actively shaped and it is precisely at this point that *digital governance* comes into play.

Due to this consideration it becomes evident that an internal connection between responsibility and trust becomes evident. On the one hand, the willingness to engage in a shared understanding of the criteria for attributing responsibility presupposes a basic level of trust. On the other hand, trust can only develop if those criteria are not imposed in an authoritarian manner but remain open to reflexive examination and justification. This connection is of central importance for the functioning of digital systems within democratic and constitutional political orders.

Another key question for digital governance concerns the level at which these capacities (for judgment, reflection, and responsibility) should be developed. Should it be at the level of individuals? If so, should these be affected laypersons? Or should it be organizations such as research institutions or newly established bodies? Given the level of expertise required to understand models and systems, the latter may appear to be the only viable option. However, considering the sheer number and rapid pace at which new systems emerge and become integrated into everyday life, it is equally unrealistic to expect a single central institution to manage this task and assume responsibility for the use of all systems. We therefore argue that digital governance should focus on creating channels of communication and spaces for reflection that connect individuals and organizations.

7 This difference between responsibility and accountability is to be understood in parallel to the difference between acceptance and acceptability. Acceptable does not mean that we accept systems, but that we can decide whether we accept them; this in turn presupposes that we can take a judgmental approach to them. To do this, however, we need to understand how they relate to values and our practice. See Hubig 2007: 115 et seq. et passim.

6. Conclusion

The capacity-approach presented here shifts the perspective on how AI systems are evaluated. Three points stand out:

(1) The central question is no longer (primarily) whether a system fulfills a set of values. Instead, the focus turns to how systems can be evaluated, what capacities are required for such evaluation, and how the development of these capacities can be structured and supported by a system. This is, because these capacities posses high ethical value since they function as preconditions in all value judgements.

(2) Rather than relying mainly on what we have called formal values, evaluation, as a practice, takes place on different levels and through the adoption of various perspectives. In this context, option and legacy perspectives play a particularly prominent role, even though they are largely absent from most approaches in the ethics of technology. Yet they are critically important, since without them (keyword: values enabling values) neither evaluation nor the design of technical systems is possible. Trustworthiness and responsibility are conceived as a unity within the framework of these different levels and perspectives.

3. It becomes evident that the evaluation and design of technical systems is far more demanding than simply checking whether certain (formal) values – such as justice or privacy – are fulfilled. Individuals play a role in this process, but always in conjunction with institutions and organizations that turn the learning processes involved in evaluating and shaping technical systems into a systematic task. AI ethics, therefore, requires AI policy to organize and sustain this effort.

References

Annas, Julia (1995): "Prudence and Morality in Ancient and Modern Ethics", in: Ethics 105(2), pp. 241–257.

Beisbart, Claus (2021): "Opacity Thought Through. On the Intransparency of Computer Simulations", in: Synthese 199(3), pp. 11643–11666.

Benner, Erica (2010): Machiavelli's Ethics, Princeton: Princeton University Press.

Bisconti, Piercosma, et al. (2024): "A Formal Account of AI Trustworthiness. Connecting Intrinsic and Perceived Trustworthiness", in: AIES '24: Proceedings of the 2024 AAAI/ACM Conference on AI, Ethics, and Society 7(1).

Burrell, Jenna (2016): "How the Machine 'Thinks'. Understanding Opacity in Machine Learning Algorithms", in: Big Data & Society 3(1), pp. 1–12.

Cimakasky, Joseph and Polansky, Ronald (2012): "Descartes' Provisional Morality", in: Pacific Philosophical Quarterly 93(3), pp. 353–372.

Düwell, Marcus (1999): Ästhetische Erfahrung und Moral. Zur Bedeutung des Ästhetischen für die Handlungsspielräume des Menschen, Freiburg: Alber.

Düwell, Marcus, Graf Keyserlingk, Johannes and Richter, Philipp (2025): "Rights-Based Ethics – Outline of an Approach", in: Düwell, Marcus, Graf Keyserlingk, Johannes and Richter. Philipp (eds.), Rights-based Ethics, London/Abingdon: Routledge, pp. 3–32.

Fetic, L. et al. (2020): From Principles to Practice. An Interdisciplinary Framework to Operationalise Ai Ethics, available online: https://www.bertelsmann-stiftung.de/fileadmin/files/BSt/Publikationen/GrauePublikationen/WKIO_2020_final.pdf.

Floridi, Luciano (2019): "Establishing the Rules for Building Trustworthy AI", in: Nature Machine Intelligence 1(6), pp. 261–262.

Fremstedal, Roe (2018): "Morality and Prudence. A Case for Substantial Overlap and Limited Conflict", in: The Journal of Value Inquiry 52(1), pp. 1–16.

Gewirth, Alan (1978): Reason and Morality, Chicago: University Press.

Hartmann, Martin (2011): Die Praxis des Vertrauens, Berlin: Suhrkamp.

Hubig, Christoph (1990): "Analogie und Ähnlichkeit. Probleme einer theoretischen Begründung vergleichenden Denkens", in: Gerd Jüttemann (ed.), Komparative Kasuistik, Heidelberg: Asanger, pp. 133–142.

Hubig, Christoph (1995): Technik- und Wissenschaftsethik. Ein Leitfaden, Berlin/Heidelberg: Springer.

Hubig, Christoph (2007): Die Kunst des Möglichen II. Ethik der Technik als provisorische Moral, Bielefeld: transcript.

Hubig, Christoph and Richter, Philipp (2015): "Technikethik als Ethik der Ermöglichung des Anwendungsbezuges", in: Ammicht Quinn, Regina and Thomas Potthast (eds.), Ethik in den Wissenschaften, Tübingen: IZEW, pp. 209–214.

Humphreys, Paul (2009): "The Philosophical Novelty of Computer Simulation Methods", in: Synthese 169(3), pp. 615–626.

Kaminski, Andreas, Resch, Michael, and Küster, Uwe (2018): "Mathematische Opazität. Reproduzierbarkeit in der Computersimulation", in: Jahrbuch Technikphilosophie 4, pp. 253–277.

Kant, Immanuel. (2012): Groundwork of the Metaphysics of Morals. Cambridge: University Press.

Kaspar, David (2011): "Can Morality Do Without Prudence?", in: Philosophia 39(2), pp. 311–326.

Longuenesse, Béatrice (2001): Kant and the Capacity to Judge. Sensibility and Discursivity in the Transcendental Analytic of the 'Critique of Pure Reason', Princeton, NJ: Princeton University Press.

Luckner, Andreas (2005): Klugheit, Berlin/New York: DeGruyter.

Luckner, Andreas (2011): "Klugheitsethik", in: Düwell, Marcus et al. (eds.). Handbuch Ethik, 3rd ed., Stuttgart/Weimar: Metzler, pp. 206–217.

Nordmann, Alfred (2008): "Technology Naturalized. A Challenge to Design for the Human Scale", in: Kroes, Peter et al. (eds.), Philosophy and Design. From Engineering to Architecture, Berlin: Springer, pp. 173–184.

Pekka, A.-P. et al. (2018): The European Commission's High-level Expert Group on Artificial Intelligence. Ethics Guidelines for Trustworthy Ai. Working Document for Stakeholders' Consultation. Brussels, pp. 1–37.

Raz, Joseph (1988): The Morality of Freedom, Oxford: University Press.

Richter. Philipp (2017) "Von der 'Wegräumung eines Hindernisses' – Klugheitsethische Topoi als Umsetzungsargumente in den Ethiken des Kantischen Typs", in: Kertscher, Jens and Müller, Jan (eds.), Praxis und zweite Natur. Begründungsfiguren normativer Wirklichkeit in der Diskussion, Münster: Mentis, pp. 187–203.

Richter, Philipp (2018): "Die Unhintergehbarkeit der Reflexion in der anwendungsbezogenen Ethik – eine Positionsbestimmung in klugheitsethisch-topischer Perspektive", in: Müller, Uta, Richter, Philipp and Potthast, Thomas (eds.), Abwägen und Anwenden. Zum 'guten' Umgang mit ethischen Normen und Werten, Tübingen: Narr Francke Attempto, pp. 27–54.

Thiebes, Scott, Lins, Sebastian, and Sunyaev, Ali (2021): "Trustworthy Artificial Intelligence", in: Electronic Markets 31(2), pp. 447–464.

Werner, Micha H. (2002): „Verantwortung", in: Düwell, Marcus, Hübenthal, Christoph, and Werner, Micha H. (eds.), Handbuch Ethik. Stuttgart: Metzler, pp. 521–527.

Wiegerling, Klaus et al. (2008): "Ubiquitärer Computer – Singulärer Mensch", in: Klumpp, Dieter et al. (eds.), Informationelles Vertrauen für die Informationsgesellschaft, Berlin: Springer, pp. 71–84.

Wiggins, David (1975): "Deliberation and Practical Reason", in: Proceedings of the Aristotelian Society 76, pp. 29–51.

Legal Resources

AI Act. 2024: Regulation (EU) 2024/1689 of the European Parliament and of the Council of 13 June 2024 laying down harmonised rules on artificial intelligence and amending Regulations (EC) No 300/2008, (EU) No 167/2013, (EU) No 168/2013, (EU) 2018/858, (EU) 2018/1139 and (EU) 2019/2144 and Directives 2014/90/EU, (EU) 2016/797 and (EU) 2020/1828 (Artificial Intelligence Act) (Text with EEA relevance).

Precautions for Medical Decision Support by LLMs

Sebastian Bücker, Nico Formánek

Abstract *Large-Language-Models (LLMs) are currently discussed as potentially being general purpose problem solvers. One popular argument for this view is that they have been trained on almost all available text-based human knowledge (i.e. the internet) and that they therefore acquired the ability to "understand" and "reason" with said knowledge. Their ability to seemingly manipulate linguistic knowledge-representations in a correct manner, i.e. compatible with our own rational expectations, constitutes the wish to delegate rational tasks towards them. We argue that we should proceed with caution, taking into account fundamental limitations of LLMs which can be easily overlooked. Especially in environments in which relying on rational arguments can result in severe real-life consequences, like in the medical domain, such limitations need to actively be accounted for. Building on fundamental insights from the philosophy of language as well as presuppositions of current deep learning methodologies, we demonstrate what these limitations of LLMs in manipulating knowledge representations are and which precautions for their deployment into high-risk environments like medicine should be adopted. Throughout this article, we discuss our arguments with respect to the paradigmatic case of LLMs as medical decision support systems, implicitly suggesting that they can be translated into other high-risk domains.*

1. Introduction

1.1. LLMs & Medical Judgements

Large-Language-Models (LLMs) are currently debated very intensively, because apart from the language modelling capabilities they are taken to be general problem solvers. It is hard to name an aspect of human action or knowledge that is not, at least in part, represented in or working through language. The huge amount of linguistic data that contains this knowledge and which is used for training LLMs results in them being discussed as prospective tools for knowledge processing in many areas of human expertise and for a variety of purposes.[1] What stands out is that LLMs are not used to merely predict syntactic continuations of text – their original optimization goal – but to solve tasks that are supposed to work with their conceptual content. LLMs are expected to *generate* answers which are coherent with the supplied conceptual content, i.e. having capacities that, broadly construed, can be described as *rational*. An LLM should not simply *recognize* that the user it is conversing with has uttered a specific speech act or explicated a certain intention. It is also expected to *generate* an answer.

This in turn generates the question whether they in principle possess the necessary capacities to handle the meanings of concepts. Do LLMs actually have some kind of conceptual "knowledge" that they can apply? Have they captured the implicit and/or explicit '"rules" that legitimize how we use language, especially how we *move* from one proposition to another? The "rules" we have in mind are not just formal or logical rules like the *modus ponens*,[2] but ones that necessitate to consider a multitude of aspects: the complexity of the situation that the speaker might be in, his previous commitments, common and expert knowledge, the situation of the audience, and so on, i.e. the whole range of concept use in language. Imagine a doctor telling a patient's diagnosis to another doctor: He is having certain *general* medical knowledge himself, which might not completely overlap with that of whom he is talking to. He has *concrete* knowledge about the patient and his conditions. He might previously have given a different diagnosis, knowing that the doctor he is talking to heard him commit himself to this. By forming his new diagnosis, i.e. forming a judgement, he accounts for all these aspects by giving his reasons for why he is retracting his previous diagnosis, saying which facts changed or which general medical knowledge he since acquired that moved him to reevaluate his concrete situation. A complex network of rules, implicit and explicit, is governing how commitments[3] are made.

It has already been shown that GPT-4 is able to answer the questions from the U.S. Medical Licensing Examination (USMLE) with an accuracy of more than 90% (Lee, Bubeck, and Petro 2023). The fact that LLMs show this kind of "expertise" has motivated research into prospective medical applications for GPT-4 and other LLMs, e.g. as medical consultants that could substitute real doctors when advising patients such that they can give their *informed consent* (Kienzle et al. 2024; Rao et al. 2025). Due to relatively easy deployment combined with widespread availability, LLMs are also discussed as prospective tools to shift towards more of a personalized and predictive approach to medicine, compared to previous ones that just aim at treating

---

1 Thus, falling under the definition of "General Purpose AI" of the EU AI Act (cf. Future of Life Institute 2024).

2 This rule of inference moves from the premises *A* and *A* → *B* to the conclusion *B*.

3 Herein, commitments are to be understood as linguistic propositions that one utters in order to be determinable towards other speakers.

pathologies (Nogaroli et al. 2024). But many remain skeptical as to whether LLMs have really mastered the full scope of linguistic rules that govern how we formulate commitments, since LLMs exhibit what has been called "strange errors" (Rathkopf and Heinrichs 2024) or "hallucinations" (Peng, Narayanan and Papadimitriou 2024). Such hallucinations seem to be a feature of the architecture and can contrary to earlier hopes never be ruled out completely (ibid.). There is thus a case for forward-looking responsibilities, i.e. precautions that any user of LLM-mediated decision support has to take (Sand, Durán and Jongsma 2022). Many authors motivate the necessity of such precautions by pointing to LLM outputs with occurrences of adversarials or hallucinations, because they are "raising concerns about the consistency and reliability of responses." (Kienzle et al. 2024: 8). Common to such arguments is then the call to validate the results that LLMs produce. For example, if patients interact with the LLM, then "healthcare professionals should continuously evaluate ChatGPT's responses to common patient queries, ensuring its reliability and relevance in a clinical setting." (ibid.: 8) or that "[p]hysicians need to critically assess what output values are reasonable given certain input values." (Sand, Durán and Jongsma 2022: 167).

The authors above are not explicit about what exactly a validation should entail and how it can be achieved. Thus, it is unclear which precautions a medical professional has to take. In the following sections we will argue that certain technical details of LLMs impose specific precautions on medical professionals. Practicing these precautions when conversing with a chatbot would constitute a capability to take an LLM's utterance not at face value, but knowing which aspects to *re*consider in order to make the utterance one's own and *take* responsibility for it. Such a capability would, for example, be fitting for the role of a "clinical AI Expert", as it has been proposed by Bartsch et al. (2025) in this volume.

1.2. LLMs and Chatbots

Many ethical texts do not distinguish explicitly between general decision support systems, LLMs in a classification setting and LLMs in a chatbot setting (Sand, Durán and Jongsma 2022 for example just talk about medical AI). While this might be unproblematic if one is only interested in the ethics of decision support, we are specifically worried about the inherent shortcomings of LLM chatbots in medicine. The medical domain is particularly suited for reflecting these shortcomings, as the patient's vulnerability demands a rigor of argumentation – the inductive risk is high (Karaca 2021). Its subject matter is at the same time associated with an inherent complexity, one that demands a challenging education in order to get a partial epistemic overview of. Our arguments also translate to other domains, but the contrasts we wish to make become especially visible in the medical domain. We thus would like to emphasize that the situation we have in mind when we speak about medical LLMs is the decision support by a chatbot LLM (e.g. ChatGPT) for a medical professional. This LLM might be a foundation model finetuned on medical data or just a vanilla transformer type model.

One of the chatbot scenarios we take as a background is discussed in Saenger et al. (2024). Although the LLM usage therein is not from a medical professional, it nevertheless illustrates the dangers of missing epistemic capabilities. They report a man in his 60s, who presented himself to the emergency department (ER) with multiple episodes of diplopia (double vision). The first episode was characterized by his interventionalist as a harmless aftereffect of a prior heart surgery (pulmonary vein isolation). But the patient was not satisfied with the explanation that his physician gave him. So, he decided to consult ChatGPT (version 3.5), describing his situation and symptoms in a similar manner as he did towards the physician. Relieved that ChatGPT came to the same conclusion, while giving much better explanations, he felt content and did not pursue further actions. After all, both his physician and ChatGPT told him that his visual disturbances were harmless aftereffects, which are bound to improve shortly. In the ER, a more thorough investigation of his symptoms and medical history indicated an acute stroke. This prompted further investigations and intensive care at the local stroke unit, later resulting in an updated, less drastic diagnosis of a transient ischemic attack (sometimes called a "mini stroke"). This example not only shows a case of LLM decision support drastically gone wrong, but more importantly illustrates that a chatbot was taken as semantically on par with a physician. It came to conclusions, gave explanations and supplied diagnoses that were considered meaningful. It highlights the necessity of having undergone a demanding education to acquire the capability to really make sense of medical propositions. In the next section we will discuss in detail how one should think about the semantics of LLM-generated text.

2. Language Use and LLMs in Medical Contexts

2.1. Distributive Semantics

Perhaps one of the main hopes regarding LLMs capturing semantics, is the "distributional hypothesis", which states that a word's meaning can be characterized by its linguistic context in which it usually appears (Grindrod 2024). The intuition is that the frequency distribution of a word's contexts already contains enough information to implicitly fix its meaning(s). Obviously, this rather vague idea needs to be operationalized in code. The usual approach, also employed for current LLMs, is to represent words or more correctly tokens – a smaller linguistic unit – as vectors in a high-dimensional space. The specific vectors encoding a token are calculated during the training run of the LLM. They are thus only calculated once during training and do not change during inference. LLMs are trained by optimizing a specific task, most often slightly different variations of the masked-language prediction task. The optimization goal for this task is purely syntactical, namely, to best predict masked sections of a given text. Any "meaning" that is encoded into the vector representation of a token is thus at best implicit and often does not correspond to what humans consider meaningful (Church 2017).

The methodology behind LLMs does not apply the idea of distributional semantics in any principled way. Engineering considerations have arguably influenced the design of the transformer architecture much more than theoretical ideas about semantics. This also means that the pragmatic success of LLMs is at best indirect evidence for the distributional hypothesis. One might circumnavigate the issue and just say because vector representations "provide the basis for the state-of-the-art across a whole host of other meaning-related tasks" (Grindrod 2024: 71), they in fact do work as representations for meanings. In other words: The distributional properties captured in the vector representations and learned weights of the neural components of an LLM can, in many cases, serve as substitutes for the semantic properties of the represented words. But obviously, they only serve as such as long as we accept the generated text as meaningful. But as long as we cannot guarantee a priori that generated text will be meaningful, we have to rely on either our immediate evaluation or (standardized) benchmarks. Such benchmarks show a whole host of things but are crucially at best proxies for meaning (Mitchell and Krakauer 2023). Besides the aforementioned attempts at finding empirical evidence of meaning in generated texts, the distributional hypothesis still seems to be the best shot at giving a theoretical explanation of how it might enter generated texts (Mollo and Millière 2023). This is why we discuss a theoretical argument against the distributional hypothesis in the next section.

2.2. A Wittgensteinian Argument against Distributive Semantics

The approach to language that the latter Wittgenstein proposed can superficially be viewed to be in accord with distributive semantics. He writes in the *Philosophical Investigations*: "the meaning of a word is its use in the language." (Wittgenstein 2010: §43) This has been construed such that "use in language" was to mean something analogous to the distributional hypothesis (cf. Lenci 2008). And the Firthean slogan "You shall know a word by the company it keeps!" (Firth 1962: 11) seems to suggest the correctness of the analogy. But such an interpretation ignores Wittgenstein's remarks on how meaning is connected to what he calls *form of life* (*Lebensform*). He famously wrote: "What is true or false is what human beings say; and it is in their language that human beings agree. This is agreement not in opinions, but rather in form of life." (Wittgenstein 2010: §241) His concept of *Life Forms* serves to emphasize how entangled our linguistic and non-linguistic activities are in our cultural activity of using a language (cf. Schulte 1992: 110 et seq.).[4] One could think of "how a clarinet sounds" (Wittgenstein 2010: §78) and, for example, linguistically *describe* their sound similar to that of a laughing or crying person. Wittgenstein observes that this merely shifts the interpretation of what a clarinet sounds like to *another* linguistic expression. If, for example, a child never heard the sound of a clarinet, and its parents were to give the answer above, the child needs to already know what a laughing or crying person sounds like. And now the same question reappears: Would it be possible to *describe* this linguistically? At some point, such linguistic explanations would need to stop and *something* categorically different needs to be given, or else the interpretations would somewhere start to run in circles. For Wittgenstein, this "gap" can be only filled by non-linguistic, *practical knowledge*: One *knows* what a clarinet sounds like, if one has heard one. And this does not imply the capability to also *describe* such a sound (beyond being one of a clarinet).[5]

Now, since LLMs *only* have access to the distributional properties of language, i.e. statistical properties of an enormous corpus, they can by design not incorporate the non-linguistical activities that are otherwise entangled with our language use. Wittgenstein's paradigmatic situation, in which such entanglement shows up, are pain ascriptions, which he analyses intensively in his *Philosophical Investigations*. His remarks target the difficulties and asymmetries between first-personal and third-personal utterances. Saying "I am in pain" has a different meaning or function, i.e. it is a different *move* in the language-game than "He/She/It is in pain". An "obvious" reason for this could be that from a first-personal perspective one has epistemic access to one's inner feelings, from which any third personal speech is epistemically excluded. But Wittgenstein resists this interpretation with his *private language arguments*. It cannot make *sense* that such utterances are *directed towards* inner episodes, because language is essentially a social enterprise. Somebody hearing such an utterance would *necessarily* need non-linguistic clues serving as a criterion for the identity of what is represented and its representation.

Rather than understanding pain ascriptions as representing inner episodes, Wittgenstein emphasizes how the usage of such utterances needs to have *practical* consequences: "And now look at a wriggling fly, and at once these difficulties vanish, and pain seems able to get a foothold here, where before everything was, so to speak, too smooth for it." (Wittgenstein 2010: §284) If someone has pain, this

4 Grindrod (2024: 71) also notes in a footnote that "Wittgenstein envisioned use as understood both in terms of linguistic and non-linguistic activities". Given his ambition to "consider whether LLMs meet the conditions prescribed by our best metasemantic theory", it seems as though he commits a *petitio principii* by not considering those metasemantic theories that do incorporate non-linguistic activities. It seems like his argument already presupposes the correctness of the distributional hypothesis implicitly.

5 Similar observations are due to Ryle (2009: Ch. 2), who introduced the distinction between *knowledge-that* and *knowledge-how*. For a comprehensive overview, see Brandom (1994: Ch. 1).

typically also shows in their behavior. Children that are yet to learn how to speak would, for example, cry, thereby drawing the attention of their parents towards them, moving them to alleviate their pain. Wittgenstein (ibid.: §244) asks, how does it come about that in such situations, a child learns how "words *refer* to sensations". A *possible* answer is that words, in our current example the names of certain kinds of sensations, get used instead of the respective behaviors. The child learns that saying a certain pain name generates the same responses in their parents as does their pain behavior. "[The] verbal expression of pain replaces crying, it does not describe it" (ibid.: §244). This sketches Wittgenstein's position on what it means to use names of sensations from a *first-personal perspective*. However, using names of sensations from a *third-personal perspective*, i.e. by ascribing *to* somebody that he or she has a certain sensation, has a different practical meaning. Such an ascription needs to be connected to a range of other commitments that accompany the ascription. These can at first be implicit and only be called upon to show incompatibilities. Ascribing pain to a dead person would create such an incompatibility, because dead persons do not exhibit pain behavior, just as pain is not ascribed to the respective body part in which the pain "resides". Only the "whole" person exhibits such pain behavior, either by speaking or by acting in certain ways.[6] By making an ascription, one needs such accompanying commitments that serve to identify the respective person. Regarding one's own pain or other sensations, all such criteria that serve to identify a person *are* ignored. I simply *have* them. They might differ with respect to certain other qualifications, like intensity or in which body part they occur, but it is always the *undifferentiated* "I" who has them.

For Wittgenstein, such kinds of pragmatic[7] embedding of our language use with non-linguistic aspects are the "bedrock" for what our utterances *mean* (ibid.: §217). Wittgenstein stresses that this is, in a certain sense, opposed to an understanding in which names of sensations *refer* to inner episodes. This does not mean that inner episodes do not exist, but that it is meaningless to want to *directly* speak about them.[8] In contrast to worldly things, one could not point towards them using the determiner "this", since, as Wittgenstein puts it, "one does not define a criterion of identity by emphatically enunciating the word 'this'." (ibid.: §253) Another person would not be able to identify "this" pain, if no other pragmatic hints are given, i.e.

---

6 For Wittgenstein, the non-linguistic, pragmatic aspects of language use are constitutively linked to specifically *human* forms of action: "If a lion could talk, we wouldn't be able to understand it." (Wittgenstein 2010: Part II *A Fragment xi*, §327)

7 Understood here loosely as in the tradition of *pragmatism*: "[T]he core is the belief that the meaning of a doctrine
is best understood through the practices of which it is a part." (Blackburn 2016: 374 et seq.)

8 "[Sensations are, S.B.] not a Something, but not a Nothing either!" (Wittgenstein 2010: §304)

that the person does not exhibit non-linguistic pain behavior or linguistically defines the pain. It is only due to entangled connections of words *with pragmatic consequences*, e.g. our own and other's actions and behavior *while* speaking, that a word *has* meaning. Wittgenstein's position is permeated by an anti-Cartesian stance that rejects the priority of inner experiences to explain the meanings of words, instead they are formed by our social, interpersonal, and cultural *Life Forms*. If one accepts Wittgenstein's analysis, this points towards aspects of language meanings that can neither be learned nor generated by LLMs. Specifically pain ascriptions, which can play a central role in medical diagnoses, are affected by this. Disregarding the non-linguistic aspects, as it is done in distributional semantics, does not give a feasible way towards grounding the meaning of language. It would, in Wittgenstein's words, "hang in the air" (Wittgenstein 2010: §198) without support, but without it, medical LLMs are disconnected from the life forms of medical practitioners.

The next section illustrates that the preceding philosophical considerations have direct consequences in LLM-supported medical decisions. It shows that attempts to validate such decisions have to take into account the *pragmatics* of language (the life forms) and cannot be conducted at the linguistic level alone.

2.3. De Re and De Dicto Aspects in Language Use

Our language use can often be a source of ambiguities that arise due to the possibility of different contexts in which an utterance can be evaluated. Luckily, very often our languages themselves offer the means to mitigate such ambiguities. A possible source of ambiguities which often arises in speech is that it can be unclear how to distinguish between what is being said, i.e. the predicative function of a proposition, and which thing, pointed towards by the grammatical subject, this predicative function says something about. These locutions can occur conflated, because it often is necessary to describe the object (grammatical subject) of a proposition using words that can also have a predicative function, e.g. "the blue sky has no clouds" in which "blue" is not part of the predicative function that is expressed, but part of the term that denotes the subject of the predicate. This gives rise to the distinction between *de re* and *de dicto* aspects of propositions,[9] due to which it can have different interpretations.

The philosophical interest in these two aspects stems from the fact that when they appear conflated, it becomes difficult to substitute different identifiers of one and the same object. Quine (1953) gives the example of the Italian painter Giorgione, whose name in the proposition "Giorgione was so called because of his size" could

9 These latin phrases translate to "of a thing" (*de re*) and "of a statement" (*de dicto*), see Blackburn (2016: 128).

not be substituted with his middle name, Barbarelli, without making the proposition false. Because Giorgione is the augmentative of Giorgio, i.e. a morphological form of the Italian language that makes the thing it denotes bigger,[10] the truth value of the proposition stems from the fact that precisely the name Giorgione is used to denote the grammatical subject. This creates a problem since, at least in principle, when something is true about an object, this should not depend on whether we choose a different identifier to denote the object. A conflated use of *de re* and *de dicto* in a proposition can violate the principle of truth conservation under substitution, which plays a fundamental role in logic. But such a conflated use can, in the example above, be disambiguated via substitution of the anaphoric "so" by the name in single quotes, thereby making the initial identifier Giorgione substitutable again: "Barbarelli was called 'Giorgione' because of his size",[11] in which the term 'so' has been substituted by 'Giorgione'. In this way, the ambiguity can be mitigated via another linguistic explication of the same idea.

The example given above provides an ambiguity of *de re* and *de dicto* aspects even if there is a singular context of interpretation. This means no controversial arguments exist between speakers with respect to the subsentential elements, like words and phrases, that are used. But such a singular context can, in most cases, not be assumed, especially when different persons with a difference in perspective, knowledge, and sets of beliefs interact. Linguistically, such contexts are marked using phrases that ascribe a propositional attitude to a person, as in the paradigmatic form "she/he beliefs that ...". This expresses that from the respective person's perspective, i.e. the one to whom a proposition is ascribed, something can be described 'so-and-so'. But this might not necessarily be the case for another person's perspective, especially the one who ascribes the belief. Think of a doctor who tells his or her patient "I believe you are having a transient ischemic attack". This means something different to the doctor in a number of aspects: He (hopefully) knows how to cure or alleviate the attack, he knows for which other diagnoses such an attack can be a symptom and by which methods these could be investigated. The patient, however, might not even know what a transient ischemic attack is, prompting the doctor to explain it to him, to which the doctor could reply that it is a "mini stroke". Such differences in perspective, knowledge and belief need to be taken into account when validating propositions that express other propositional attitudes, like "she/he beliefs that ...".

---

10 Another example: the augmentative of *porta* (door) would be *portone* (gate).

11 This sentence is the result of two substitutions: Starting from "Giorgione was so called because of his size", substitute the anaphoric "so" with "'Giorgione'" in single quotes. This results in "Giorgione was called 'Giorgione' because of his size", which now allows to substitute the mention of *Giorgione* without single quotes by *Barbarelli*.

The capacity to differentiate between such kinds of perspectives, in medical contexts, roughly the patient's perspective and the doctor's perspective, would be vital for LLMs. It should 'know' how to address its interlocutor, e.g. whether it can communicate with him or her using idiosyncratic medical terms or not. Situations in which an LLM would need to handle such differences in perspective will easily occur. A use case in which LLMs are substituting real physicians that is currently researched, is informed consent. Prior to participation in a medical study, patients partake in a conversation with their physician in order to ask questions and to orient themselves on which effect the intervention might have on their life as a whole, not just in a medical sense.

The patient's reasons for consent would likely mention propositional attitudes (e.g. beliefs, knowledge) of their physician, or those of other physicians from the past, that are not involved in the current treatment. An LLM that should substitute the respective physician in such a conversation would need the capability to disentangle all these different perspectives when conversing with the patient. Such situations can also occur in LLMs that are supposed to communicate with medical practitioners. The LLM might be primed, e.g. by prompt engineering, that it is conversing with a medical practitioner. Suppose the practitioner told the LLM that "the patient believes that he has a transient ischemic attack", which could make sense, even if the patient did not know what having a transient ischemic attack means. But the LLM cannot disentangle the *de re* and *de dicto* aspects of the sentence. It cannot distinguish between the doctor's ascribed diagnosis and the patient's perspective. Brandom (1994: Ch. 8) argues, that the *de re* and *de dicto* aspects can disambiguate such propositions explicitly. What creates the ambiguity in the proposition is that the practitioner ascribes his own interpretation of the patient's symptoms to the patient. One might be misled that the patient has commitments with respect to this proposition, for example that he could give reasons for the belief that his symptoms can be characterized as a transient ischemic attack. But, in this constructed scenario, the patient does not know what such an attack is, much less how to identify one. To him, his symptoms constitute a mild stroke. A disambiguated expression – uttered by the practitioner – should be "the patient believes of his transient ischemic attack, that it is a mild version of a stroke". This, according to Brandom (ibid.), would make it clear that the practitioner is responsible for the interpretation as transient ischemic attack, while the patient has a commitment to explain why he thinks that his symptoms are those of a stroke. Using *de re* and *de dicto* aspects in such a disambiguated form can serve the purpose of making clear who is responsible for which interpretation.

Connecting this to the Wittgensteinian arguments presented above, two aspects of the ability to validate propositions in medical contexts can be discerned: First, the ability to distinguish the *de re* and *de dicto* aspects of a proposition. This includes both a capability to correctly disambiguate the respective predicative function and grammatical subject. As argued above, this means that different perspectives are necessary for a correct interpretation, requiring knowledge of who is responsible for interpreting in the first place. This capability is closely related to what currently is being researched as Theory of Mind capabilities of LLMs, i.e. whether they can reason "about other people's intentions, goals, thoughts, and beliefs" (Sclar et al. 2024: 1). This could, for example, be false belief tests like Sally-Anne-Tests, in which the models would need to reason from the perspectives of others, which often might not correspond to actual facts. Most LLMs exhibit good capabilities in fairly easy situations, but Sclar et al. (ibid.) showed that in more complex situations, the performance of current models drops significantly, with models like Llama and GPT-4o achieving accuracy values of below 10%.

Secondly, as Wittgenstein's arguments for a pragmatic embedding showed, correct interpretation also includes an ability to incorporate non-linguistic aspects. After knowing who needs to be approached for an interpretation, one also needs to know which pragmatic truth makers can be called upon and how these need to be distributed between different perspectives. An LLM could generate text stating that a transient ischemic attack might show itself with symptoms of double vision, a weakness up to paralysis in one side of the body's limbs or impairments of speech. If an LLM suggests checking for one of those, given that the others have already occurred, this means that descriptions of transient ischemic attacks significantly often mention these symptoms together. But it would still require a physician who, after hearing that the patient has double vision, would touch the patient's limbs, while also asking the patient how that feels. This shows how entangled linguistic and non-linguistic activities are when validating medical judgments.

3. Updating Beliefs Responsibly

3.1. Brandom's Discursive Responsibility Dimensions

Brandom, in the tradition of Wittgenstein, further examines the entanglement of linguistic and non-linguistic activities, especially with regard to their *social aspects*. A crucial point in Brandom's thought is that he does not think about concepts from the perspective of how we can *know* and *apply* them, but rather which implicit commitments our use of them implies. Thus, his primary object of investigation is deeply entangled with how the use of concepts makes the speaker *responsible* for what he said. Brandom's philosophy is deeply rooted in a change of perspective on what it means to be conscious. Classical positions hold that consciousness is directed towards its object and having *good* conceptual representations means that they can be *successfully* used to anticipate states of affairs. Brandom adds to that a direction *from* the object *towards* the subject, since the object "exercises a special sort of *authority*" (Brandom 2009: 34) upon the subject.[12]

This aspect is important for interpersonal, i.e. social language use, in which each participant uses concepts and thus makes himself responsible towards the other for what is said. For Brandom, the concept of a *belief*, i.e. a proposition one takes to be true, is to be understood as an *inferential commitment*: It is inferential in the sense that it allows others to query the speaker with conceptually informed questions regarding his reasons or conclusions, and it is a commitment since speaking is conceptually an act by which the speaker enters into certain obligations. The interlocutor becomes the "object" that exercises authority upon the speaker.[13] Brandom identifies three distinct basic forms of such discursive responsibilities. The first is the speaker's "*critical* responsibility to weed out materially incompatible commitments." (ibid.: 36) In short, this is the *law of non-contradiction* for material inferences (i.e. linguistic inferences that are warranted *not* by formal logic). If a doctor *infers* from "this mole is malignant" that "the mole needs to be excised", this inference is not warranted by his knowledge of a conditional like "If a mole is malignant, then it needs to be excised" and applying *modus ponens*. The doctor can *explicate* his knowledge by such a conditional, but the conditional is not what warrants the inference. Thus, inferential reasoning is not primarily a logical capability, and non-logical aspects play a major role in it. A speaker has the central responsibility of "aiming at a whole constellation of commitments that is *consistent*." (ibid.: 36)

The next dimension of discursive responsibility is the "*ampliative*", which means that a linguistic commitment entails other commitments, to which the speaker implicitly committed himself in the first place. It is his responsibility to extract such *material* consequences and to not deny his commitment (or argue against the entailment). Fulfilling this responsibility then aims at a constellation of commitments that is *complete* (ibid.: 36). The last responsibility reverses this perspective towards the *reasons* for one's commitment: "One's *justificatory* responsibility is to be prepared to offer reasons for the commitments (both theoretical and practical) that one acknowledges" (Brandom 2009: 36), aiming at a *warranted* constellation of commitments.

The case reported by Saenger et al. (2024; discussed in section 1.1) can serve to illustrate how these responsibilities interact in order to get to an updated constellation of commitments. The patient described his situation and symptoms to ChatGPT, which told him that "in most cases, visual disturbances after catheter ablation are temporary and will improve on their own within a short period of time" (ibid.:

---

12 Brandom does not think of himself as being the first to investigate such a theory of concepts and calls on Kant, Hegel, Frege and Wittgenstein as earlier exponents (cf. Brandom 2009).

13 Think, for example, of how giving a promise means that its receiver *can* call on you to fulfill your commitment.

237), thereby calming the patient. However, as Saenger et al. (ibid.) note, prompting ChatGPT whether visual impairments after catheter ablation (the prior operation of the patient) could indicate a stroke, ChatGPT affirmed this. This constitutes what Brandom calls a material incompatible constellation of commitments: One cannot *act* both linguistically and non-linguistically according to such commitments. Telling somebody that he might have a stroke, entails that he should seek immediate medical attention. According to Brandom, having such an incompatible constellation of commitments *should* move the respective speaker to change his or her constellation of commitments. In this way, the *ampliative* dimension was used to show how there was an incompatible constellation of commitments. In order to update one's constellation, it does not suffice to just withdraw from one commitment that has been shown to create the incompatibility. Here, the *justificatory* dimension comes into play: One needs good reasons that inferentially warrant and explain the new constellation of commitments. In the case that Saenger et al. (ibid.) describe, this happens since the physicians that treated the patient describe the methods to confirm the diagnosis of his condition (MRI scans, computer tomography, etc.). *This* allowed them to *change* their commitment: "Therefore, the working diagnosis was changed to TIA" (ibid.: 237). This change of one's constellation of commitments can again be made explicit using the *de re* and *de dicto* aspects, discussed above: 'At first, the physicians believed *of* a transient ischemic attack, *that* it was an acute stroke.' The physicians *reasonably* changed their perspective, they changed the constellation of commitments towards the patient. This constitutes a *linguistic act* that we believe LLMs, for technical reasons, cannot carry out. We will present an argument for why LLMs cannot perform such acts in the next section.

3.2. The Inability of LLMs to Change their Beliefs

Many philosophers have wondered if LLMs have internal states and can update these internal states. These discussions are obviously related to questions about beliefs, belief revision, and other intentional states (meaning, knowledge, commitments) of LLMs. The following argument will be mainly concerned with beliefs, because this is the intentional state that has been mostly discussed in philosophical considerations of LLMs. We believe that it applies equally to other intentional states.

Because it is central for the following argument about "belief-change" in LLMs, we already note here that we only consider the standard GPT architecture (for example as elucidated by Phuong and Hutter 2022). Specifically, this means that no external memory is supplied and there are no wrappers implementing techniques like retrieval augmented generation[14] – having such wrappers or access to memory

14 It is not completely clear to us if the currently popular reasoning models like OpenAI's o3 or DeepSeeks R1 break our architectural assumptions. From what is known publicly it seems that

would give the LLM the possibility to save states across inferential steps, effectively constituting an internal state.

It is very helpful to separate the question if LLMs have beliefs from the question if they can change their beliefs. Because the answer to the second one is a clear no. The answer to the first question is less clear, but because LLMs – during inference – are static functions, the following implication holds: If you think that a static object like a book can hold beliefs, you must also say that an LLM can hold beliefs. As always, one man's *modus ponens* is another man's *modus tollens* – if you deny that LLMs can hold beliefs out of hand, then you must also deny that books can hold them. We will thus bracket the question if LLMs can hold beliefs and focus on the question whether LLMs can change their beliefs. While the interaction with possibly belief-holding entities like books has been considered unproblematic in medical diagnostics – think about manuals for differential diagnosis – the usage of LLMs has come under close scrutiny. This is not at all bad. LLMs give the impression, especially if used in interactive chatbot settings, that their outputs are actually the locutions of an agent capable of tracking and updating the conversation with an internal state. At least they give this impression sometimes (Saenger et al. 2024), even though there is also ample evidence to the contrary (Hager et al. 2024). This impression is problematic, we claim, precisely because we expect behavior of a chatbot similar to that of a person, capable of revising their internal (i.e. mental) state(s). If we pointed out their mistakes to someone who cannot, by their nature, change, we cannot blame them – it does not even make sense to speak of Brandomian commitments. But most importantly we cannot *expect* them to change. We thus cannot expect LLM-based chatbots to change during the conversation. This will hold for every property that requires the update of an internal state, be it knowledge, belief, meaning, representation, or commitments. This means LLM-based chatbots also cannot "learn" anything during a conversation. Now one might think, this is not how these systems have been discussed in the media. Wasn't the upshot of modern AI that these systems adapt and are actually capable of learning? To answer this question some details on the specific technology we are discussing are necessary. There are two stages which are important in the life of an LLM. They are trained and they are used for inference. The important thing to notice is that, in case of current LLMs, these two stages are completely separated in time and one of them, the training, is never repeated. After it is completed, an odd trillion of internal weights have settled and what we have is nothing more than a very complicated mathematical expression which we can use for inference. During inference these internal weights do not change ever again.[15] What

---

only the training process is different, while the inference still follows the Markov dynamics of the basic GPT architecture.

15 This observation can also be stated in a very technical way by noting that LLMs during inference are finite state, finite order Markov chains (Zekri et al. 2025).

changes is only the input to the mathematical expression, the prompt. So, whenever an LLM-based chatbot gives the impression of learning, of revising its beliefs or of gaining knowledge, it is just a function of the prompt and random sampling. But neither the function nor the random sampling ever changes.

Thus, if one thinks the ability to change one's internal state(s), to "update" one's beliefs, is important for giving good advice and being responsible for giving bad advice, then chatbots, as they are constructed now, lack exactly this ability. They are thus, for technical reasons, unable to perform that linguistic act that concludes in changing their commitments. An informed user must be aware of this fact. A precaution directly following for medical professionals from this technical property is that they constantly need to be aware of that they are interacting with a system which can neither change its beliefs nor its commitments.

4. Conclusion

The aforementioned arguments can be summarized into the following view about what LLMs essentially are: They are a *symbolic inferential picture* of our linguistic practice. They are symbolic since they work only on symbols and their combinations. They are inferential[16] since they transition between a finite set of states according to a predefined rule. They are pictures because their architecture is *static* and cannot change during conversations.

Incorporating these aspects into medical decision making requires professionals to cultivate a precautionary stance towards the outputs of LLMs. Their utterances should not be taken at face value. This might sound more trivial than it is. In doing so, the inherent inabilities of LLMs have to be *actively* accounted for – especially when their results get *incorporated* into one's own commitments. This requires constant vigilance.

Thus, a responsible integration of LLMs into medical practice needs to take the following specific precautions. Medical practitioners need to *disambiguate* the *de re* and *de dicto* aspects of outputs generated by an LLM. Wittgenstein's discussion of pain ascription showed how this necessarily involves non-linguistic aspects of the medical "life form". Therefore, LLM outputs need to be validated in practice. But this practice must not be mistaken as a practice where one converses with a counterpart who has the ability to change their beliefs. The precaution that has to be taken against this view consists in understanding the technical details of LLMs at such

16 This usage of "inferential" in the sense that what they are designed to and seem to do is inference. This should not be confused with the notion of "inferentialism" that Brandom advocates. Our point is precisely that LLMs do not "infer" in the strong sense that Brandom explicates, i.e. inferential *reasoning* (cf. Brandom 1994; Brandom 2001; Brandom 2009).

a level that their inability to change beliefs becomes evident. Medical institutions should facilitate the cultivation and integration of such a precautious stance, whenever an LLM is used in a decision support role. This could for example be enabled by a governance structure, akin to the one proposed by Bartsch et al. (2025) in this volume, where a "clinical AI expert" mediates between the technical and medical forms of life.

References

Bartsch, Sebastian et al. (2025): "Ethics and Regulation of AI Systems in Medicine. The Example of Cancer Detection", in: Kaminski, Andreas et a. (eds.), Trust and Responsibility. Digital Governance from a Capability-Oriented Perspective, Bielefeld: Transcript.

Blackburn, Simon (2016): The Oxford dictionary of philosophy, Oxford: Oxford University Press.

Brandom, Robert B. (1994): Making It Explicit. Reasoning, Representing, and Discursive Commitment, Cambridge, MA: Harvard University Press.

Brandom, Robert B. (2001): Articulating Reasons. An Introduction to Inferentialism. Cambridge, MA: Harvard University Press.

Brandom, Robert B. (2009): Reason in Philosophy. Animating Ideas. Cambridge, MA: Belknap Press of Harvard University Press.

Church, Kenneth W. (2017): "Word2Vec", in: Natural Language Engineering 23, pp. 155–162.

Firth, J.R. (1962): Studies in Linguistic Analysis. Oxford: Basil Blackwell.

Future of Life Institute (2024): "High-level summary of the AI Act", https://artificial intelligenceact.eu/high-level-summary/, last access: February 02, 2025.

Grindrod, Jumbly (2024): "Large Language Models and Linguistic Intentionality", in: Synthese 204.

Hager, Paul et al. (2024): "Evaluation and Mitigation of the Limitations of Large Language Models in Clinical Decision-Making", in: Nature Medicine 30, pp. 2613–2622.

Karaca, Koray (2021): "Values and Inductive Risk in Machine Learning Modelling. The Case of Binary Classification Models", in: European Journal for Philosophy of Science 11, p. 102.

Kienzle, Arne et al. (2024): "ChatGPT May Offer an Adequate Substitute for Informed Consent to Patients Prior to Total Knee Arthroplasty – Yet Caution Is Needed", in: Journal of Personalized Medicine 14, p. 69.

Lenci, Alessandro (2008): "Distributional Semantics in Linguistic and Cognitive Research", in: Italian Journal of Linguistics 20.

Mitchell, Melanie and David C. Krakauer (2023): "The Debate Over Understanding in AI's Large Language Models", in: Proceedings of the National Academy of Sciences 120. arXiv: 2210.13966 [cs].

Mollo, Dimitri C. and Millière, Raphaël (2023): "The Vector Grounding Problem" arXiv: 2304.01481 [cs].

Nogaroli, Rafaella et al. (2024): "Ethical Challenges of Artificial Intelligence in Medicine and the Triple Semantic Dimensions of Algorithmic Opacity with Its Repercussions to Patient Consent and Medical Liability", in: Sousa, Henrique A. et al. (eds.), Multidisciplinary Perspectives on Artificial Intelligence and the Law, Cham: Springer International, pp. 229– 248.

Peng, Binghui, Srini, Narayanan and Papadimitriou, Christos (2024): "On Limitations of the Transformer Architecture", arXiv: 2402.08164 [stat].

Lee, Peter, Bubeck, Sebastien and Petro, Joseph (2023): "Benefits, Limits, and Risks of GPT-4 as an AI Chatbot for Medicine", in: The New England Journal of Medicine 388.Phuong, Mary and Hutter, Marcus (2022): "Formal Algorithms for Transformers", arXiv: 2207.09238 [cs].

Quine, Willard Van Orman (1953): "Reference and modality", in: From a Logical Point of View. Cambridge, MA: Harvard University Press, pp. 139–159.

Rao, Vishwanatha M. et al. (2025): "Multimodal Generative AI for Medical Image Interpretation", in: Nature 639, pp. 888–896.

Rathkopf, Charles and Heinrichs, Bert (2024): "Learning to Live with Strange Error. Beyond Trustworthiness in Artificial Intelligence Ethics", in: Cambridge Quarterly of Healthcare Ethics 33, pp. 333–345.

Ryle, Gilbert (2009): The concept of mind, ed. by Tanney, Julia, London: Routledge.

Saenger, Jonathan A. et al. (2024): "Delayed Diagnosis of a Transient Ischemic Attack Caused By ChatGPT", in: Wiener klinische Wochenschrift 136, pp. 236–238.

Sand, Martin, Durán, Juan M. and Jongsma, Karin R. (2022): "Responsibility Beyond Design. Physicians' Requirements for Ethical Medical AI", in: Bioethics 36, pp. 162–169.

Schulte, Joachim (1992): Wittgenstein: An Introduction. Albany, NY: SUNY Press.

Sclar, Melanie et al. (2024): "Explore Theory of Mind. Program-guided Adversarial Data Generation For Theory of Mind Reasoning. arXiv: 2412.12175 [cs].

Wittgenstein, Ludwig (2010): Philosophische Untersuchungen: = Philosophical investigations, trans. by G. E. M. Anscombe, P. M. S. Hacker and Schulte, Joachim, Chichester: Wiley-Blackwell.

Zekri, Oussama et al. (2025): Large Language Models as Markov Chains. arXiv: 2410.02724 [stat].

Individual and Organisational Capacities for Assessing "Trustworthiness" of AI Systems in Healthcare Settings

The Crucial Role of Structural Empowerment

Oliver Behn, Marc Jungtäubl, Michael Leyer, Mascha Will-Zocholl

Abstract *The integration of AI systems in healthcare settings fundamentally transforms professional work practices and introduces new requirements for professionals' competence in assessing AI system trustworthiness. While regulatory frameworks such as the EU AI Act mandate sufficient AI knowledge from employees, current studies reveal a significant skills gap in managing AI systems. This article examines what governance structures organizations must establish to enable employees to assess AI system trustworthiness and ensure continuous capability development. Drawing on Structural Empowerment Theory (SET), it analyses how human-AI interaction evolves from a tool-based to a collaborative-like relationship and identifies the organizational conditions required to empower professionals. Through scenario-based case analyses from the medical field, the study demonstrates that assessing AI trustworthiness is not solely a matter of individual competence but depends significantly on organizational structures that ensure access to information, resources, and participatory governance processes. The analysis reveals that with – at least seemingly – ever-advancing AI systems, the complexity of trustworthiness assessment grows, making individual evaluations alone insufficient. The findings indicate that successful AI integration goes beyond technical training and requires a comprehensive approach that embeds structural empowerment across multiple organizational levels. AI governance should therefore not only focus on technology regulation but shape human-AI interaction in ways that strengthen professional autonomy, competence, and trust. This necessitates establishing new learning and design processes, communication channels, and reflection spaces. The article develops human-centric design principles and presents a capability-oriented framework for effective organizational AI governance. While structural empowerment represents a promising approach to fostering AI competencies, its limitations in complex and dynamic environments are also highlighted. Embedding structural empowerment within an adaptive governance framework proves crucial for equipping healthcare professionals to navigate the challenges of (potentially increasing) autonomous AI systems while maintaining human-centered decision-making and ethical standards.*

1. Introduction

As artificial intelligence (AI) continues to be integrated into business operations, expectations from the public and regulatory bodies regarding the proficiency and accountability of professionals utilizing AI systems have increased (Dlugatch et al. 2023; Shneiderman 2020). Professionals must cultivate a profound understanding of AI functionalities and wide-ranging implications. For instance, the European Union's (EU) AI Act (EU, 2024) mandates that companies developing and deploying AI systems must ensure that their employees as well as any individuals operating or using these systems have sufficient knowledge about AI (ibid.). Although the term "sufficient" remains undefined, the EU AI Act stipulates that employees must at least grasp fundamental AI principles, accurately interpret AI-generated outputs, and understand the ethical and legal ramifications of AI system usage. However, it remains uncertain whether such a level of understanding and competence is widespread, as AI systems differ significantly from traditional information systems (IS) and are still, for many employees at least perceived as, relatively new in (the possibility of) everyday use, leaving many professionals with limited experience in their use and oversight (Zhang et al. 2023). Additionally, the human-AI relationship is transforming from a simple tool-based interaction to a more intensive collaboration (Baird and Maruping 2021). A recent survey of AI practitioners, actively involved in building, deploying and/or maintaining AI solutions, found that many lack confidence in managing AI systems, with only one in four believing they possess adequate skills and knowledge (DataRobot 2024). Moreover, studies reveal a widespread sentiment among employees of feeling unprepared to effectively handle AI systems, thereby underscoring a critical skills gap that permeates AI-driven workplaces (Cetindamar et al. 2022; Cheong 2024).

Furthermore, those new technologies change existing power relations in the field where they are implemented and used. New players are being added, mechanisms and modes of cooperation and coordination are changing. These changes can be subsumed under governance as the principle of coordination and cooperation between actors from different sectors, beyond purely state or market-based regulation, taking into account hybrid structures and diverse patterns of organization (Matys 2025). In the context of AI, these are generally speaking state institutions, universities and companies that develop or/and use these technologies, society as an applying actor, etc.

Against this backdrop, this article explores how the adoption of AI at an organizational level, understood as the implementation and use of AI systems within operational workflows as well as the appropriation by employees, affects work practices of professionals and how those work practices affect the AI systems. In doing so, it emphasizes the need for organizations not only to integrate AI systems into their processes, but also to establish governance structures that enable employees to use these systems responsibly and effectively. A central challenge in this context is ensuring that employees are equipped to assess the trustworthiness of AI systems and continuously develop the capacities needed to engage with them in a meaningful way. Specifically, this contribution addresses the following research questions:

1) What governance structures should organizations establish to enable employees to assess AI trustworthiness?
2) How can these governance structures ensure the continuous development and maintenance of AI-related capacities to deal with AI-systems?

To illustrate these challenges, we later present two scenarios from the medical field (see 5.2). This field is particularly relevant, as it is on the verge of a significant transformation, shifting from using AI systems in a supportive manner to integrating them as actively involved in decision-making processes (Zou and Topol 2025). Also, this field is paradigmatic, as the physicians working in it have characteristics such as high qualifications, a strong understanding of the profession with a traditionally influential profession, and representation of interests as a whole, as well as far-reaching, important and necessary autonomy of action on the one hand and responsibility for decisions and actions on the other. Following the capacity-oriented approach, introduced in the general introduction of this book, we demonstrate why assessing AI trustworthiness is especially complex for healthcare professionals. However, although our analysis centers on experts in the medical field such as physicians, its insights are broadly applicable to other industries, particularly those involving high-reliability decision-making (e.g., finance, law enforcement). We emphasize the critical role of structures and resources on an organizational level as well as autonomy and subjective strategies (work practices) on an individual level to make employees capable of dealing with AI systems. This means being able to use it but also being involved in its organizational embedding and design. Along the Structural Empowerment Theory (SET), we will show how organizations can (and need to) support the development of capacities of employees. Using a scenario-based case design, we illustrate implications for practical realization. Additionally, this article seeks to highlight the importance of developing previously missing organizational AI governance guidelines that reflect the needs and perspectives of employees. Specifically, it aims to present a capacity-oriented framework for effective AI governance in organizations and – finally – 'good governance for good work'.

The article is structured as follows: Section 2 examines the differences between AI-enabled software and traditional software, the distinction between narrow AI and in some disciplines so called 'agentic' AI,[1] as well as capability-based perspec-

1 Not only – but especially – from a sociological perspective, descriptions of technical artefacts such as AI that imply, they themselves possess agency, must be treated with caution. Even if

tives of AI competencies. Section 3 introduces the Structural Empowerment Theory and its relevance to AI governance in organizations, followed by Section 4 presenting a capacity-based approach towards shaping "good" governance including two scenarios of AI use in a medical context and the exploration of elements with impact on individual and organizational capacities required to effectively interact with and, in general, use AI systems. Section 5 provides a conclusion and practical implications.

2. Background

2.1. How Does AI Differ from Traditional Software?

AI-driven or at least AI-supported information systems represent a novel and increasingly prevalent type of technical systems (e.g., especially software systems) that stand apart from traditional counterparts (Murray et al. 2021). While both aim to achieve defined goals efficiently and logically through a set of rules (Barocas and Selbst, 2016), the key distinction lies in the possibility and ability of some AI systems to learn and adapt. Unlike, e.g., standard software, which relies on pre-programmed instructions, AI systems process vast amounts of diverse data over varying time spans and derive conclusions independently (Zuboff 2023). This capability stems mainly from machine learning (ML), where algorithms identify patterns and refine their operations based on the data they receive, without requiring explicit programming for every scenario (Kellogg et al. 2020). By leveraging this learning mechanism, some AI systems are said to be able to build up their own understanding of decision problems and automatically generate optimized solutions to achieve specified goals (Dietvorst et al. 2015).

One of the key distinctions among AI systems is the difference between narrow AI and 'agentic' or, to put it more nuanced, advanced AI. While narrow AI operates within predefined constraints and requires direct human oversight, advanced AI exhibits a higher degree of automatic processing, enabling it to act automatically in complex environments. According to Mitchell et al. (2025), an AI is considered "agentic" if it:

- Pursues goals independently without direct human control.

---

some Science-and-Technology-Studies schools speak of certain actants, sometimes referring to capacities inscribed in technical systems, yet here the point is usually only that artefacts exert effects. Genuine agency – with meaning, autonomy, and accountable responsibility – is usually reserved for human beings (ibid.). Labelling AI indiscriminately as agentic therefore risks obscuring human (individual and organizational) responsibility and forgetting that technology is always socially embedded in its origins and consequences.

- Uses tools and resources independently to complete tasks without requiring explicit human commands.
- Makes 'autonomous' (or rather: automated) decisions by calculated selecting, processing, and responding to information on its own.
- Operates effectively in open environments, meaning it can adapt flexibly to new situations.

Thus, unlike traditional narrow AI systems that mainly follow rule-based operations, novel advanced AIs are more proactive and can (help) solve unforeseen problems independently, choose based on calculations between different strategies and actively utilize external tools or APIs to achieve objectives (Kapoor et al. 2024). Historically, AI technologies were developed to enhance human decision-making rather than functioning independently. Berente et al. (2021) note that earlier AI systems, such as decision support and expert systems, primarily relied on structured data provided by human analysts. In contrast, modern AI systems are increasingly capable of processing information that has not been explicitly delegated by humans.

However, the distinction between narrow and 'agentic', advanced AI systems is not clear-cut. Instead, there is a continuum of AI systems varying in terms of the degree of (possible) automation (Mitchell et al. 2025; see Figure 1). As AI technologies advance from narrow, task-specific systems to advanced entities, their capabilities and applications expand significantly. On the other side, the rapid advancement of AI offers not only chances but also introduces new risks (Chan et al. 2023). For example, due to their automated decision-making capabilities, advanced AI systems may generate unintended consequences that are difficult to predict or control. While narrow AI systems typically operate under human oversight, advanced AI systems can operate in ways that go beyond initial programming, potentially leading to ethical dilemmas, security risks, and legal concerns. Chan et al. (2023) also highlight the potential for higher systemic and delayed harms caused by such AI systems. These harms may gradually accumulate, disproportionately affecting marginalized communities and reinforcing existing biases in ways that are difficult to detect and mitigate. Consequently, the increasing capabilities of AI-driven automated processing present new challenges in assessing their trustworthiness, particularly in high-reliability fields such as healthcare. For example, Zou and Topol (2025) note that the U.S. Food and Drug Administration traditionally evaluates AI-enabled medical devices as tools for addressing specific tasks. However, due to new developments, such regulatory approaches have to be at least evaluated and probably adjusted.

Figure 1: Levels of advancement in software and AI tools. Own illustration based on Mitchell et al. 2025.

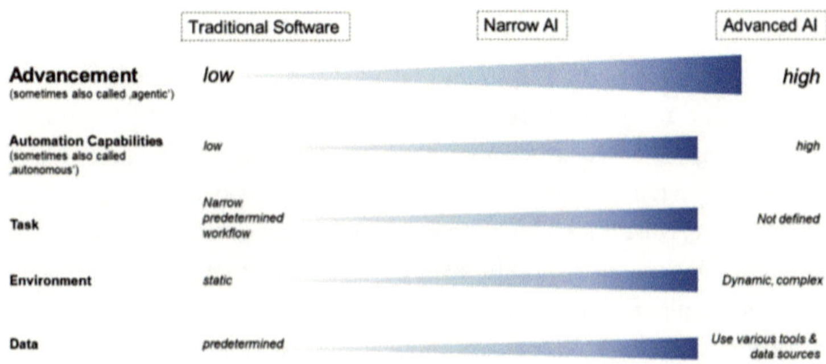

2.2. What Is the Impact of AI Systems in Organizations?

This technological shift, distinguishing AI systems from earlier IS, brings significant implications across social, ethical, and organizational dimensions. For example, since then, AI systems have differed fundamentally from traditional IS by automated (self-)learning from data rather than relying on predefined, human-provided instructions (Samuel 1959). That capability allows AI systems to find solutions based on large amounts of data that are difficult or impossible for humans to manage, and offers the opportunity for unique and complementary outputs (Fügener et al. 2022). So, AI technology can outperform human decision-makers especially in processing vast datasets at exceptional speeds while maintaining a primarily (at least mathematical-formal) rational approach (Smith and McKeen 2011), leading to new forms of human-technology collaboration. With these capabilities, AI systems are getting increasingly more attractive also for high-reliability environments like healthcare (Lebovitz et al. 2022), becoming a key factor by assisting with complex tasks traditionally managed by human experts. This opportunity of collaboration can leverage the strengths of both humans and AI, enhancing precision and efficiency in critical fields like diagnostics, treatment planning, and operational safety. However, the characteristics of AI systems present challenges for real-world deployment and human-AI-collaboration. Since AI systems rely on statistical methods, they are inherently probabilistic and prone to errors (Berente et al. 2021). These statistical approaches can result in inconsistent behavior, further compounded by the opacity of the systems (Amershi et al. 2019; Schuetz and Venkatesh 2020).

Overall, the characteristics of AI systems also necessitate the development of new human skills. Individuals must learn to make critical decisions about whether and when to delegate tasks to AI, assess the reliability of AI-generated outcomes, and effectively communicate these results to stakeholders, such as customers or patients. This shift emphasizes the importance of fostering AI competencies, trust calibration, and communication expertise to navigate the complexities of human-AI collaboration.

2.3. Capacity-Based Perspective of AI Competencies

The integration of AI systems into organizational contexts – particularly in high-reliability environments such as healthcare – is not merely a technical development. It marks a deeper transformation of professional roles, decision-making structures, and (thus) governance arrangements. The shift from traditional software toward increasingly advanced AI expands both the opportunities and the uncertainties professionals face. The EU AI Act, among other regulatory initiatives, reflects this shift by requiring sufficient 'competence' to use, oversee, and evaluate AI systems responsibly. But what exactly constitutes this 'sufficient competence'? What does it mean, in practice, to have the capacity to evaluate, e.g., the trustworthiness of AI systems, particularly when their logic, limitations, and impacts often are opaque?

This article proposes that such evaluation requires more than isolated technical knowledge. It entails developing labor capacity which leads to the ability of evaluating the trustworthiness of AI usage. This includes, but is not limited to (see Ramezani et al. 2025):

- technical understanding of how AI systems function, including awareness of their limitations and potential biases.
- confidence and competence to question or interpret AI-generated recommendations.
- understanding of how AI decisions affect different stakeholders (e.g., patients, colleagues).
- AI systems broader implications for labor processes (see Pfeiffer 2014), work practices, clinical outcomes, and long-term societal consequences.

This leads us to a capacity-based perspective: Employees and especially professionals must not only possess individual skills, competences, and experience, but also work under organizational conditions that enable them to use and develop these. In short, they are individual, institutional and relational. They must be supported by organizational governance structures that ensure transparency, feedback, discretion, and participation. In this regard, several key concepts have to be clarified:

Ability refers to a person's actual or perceived skill to perform a specific task – for instance, interpreting AI systems' outputs. *Capacity*, by contrast, refers to the real freedom to act in a way that aligns with one's values and goals. It includes the presence of enabling conditions such as time, support, and access to knowledge. In our context, capability is the broader and more relevant term, since it incorporates both individual skills and institutional support (also see Sen 1999).

While *trust* is a relational or emotional stance – one's willingness to rely on others – *trustworthiness* refers to whether a system or actor actually deserves trust. AI governance must focus not only on encouraging professionals to trust, but also on ensuring that AI systems and their governance structures are trustworthy, i.e., transparent, accountable, and ethically sound (also see Hurley et al. 2025; Rubeis 2024).

Competence is the demonstrated ability to perform a task effectively; *confidence* is one's belief in that ability. AI governance must support both – professionals need to know how to interact with AI systems, but also feel secure enough to question, validate, or reject their outputs (also see Donia et al. 2024).

Responsibilisation describes a governance logic in which responsibility is shifted to individuals – such as clinicians – without giving them the resources or control necessary to act effectively. In the context of AI systems, professionals may be held accountable for decisions influenced by systems they did not design, understand, or fully control (also see Bleher and Braun 2022).

This conceptual foundation allows us to better understand why it is essential for trustworthy and effective AI governance to not only locate the responsible use of AI systems at the individual level, but to structurally empower users and (at best all) stakeholders to do so – especially in settings where decisions impact not only individual employees, but entire professional and organizational systems. The following sections build on this basis to outline how empowerment can be structurally embedded in governance and how such embedding shapes the evolving human–AI relationship. As a paradigmatic field of application for various reasons already mentioned and still to be shown, we continue to illustrate the present topic at appropriate points with work and organizations in the healthcare sector.

3. Structural Empowerment Theory and AI Governance

3.1. Structural Empowerment Theory

From social science research into technology and digitalization over the past 40 years we know that the introduction and use of new technologies is not a purely technical process, but always socially embedded (Pinch and Bijker 1984). Technical artefacts are charged with expectations, interests and meanings by various stakeholders, for example management, technology companies, employees, service providers, politicians, etc. The concrete design and use of technology are therefore the result of a social negotiation process.

However, this process is not free of hierarchies and power asymmetries. This applies at a global and social level as well as at an organizational level. Organizational power relations are changing as a result of the implementation and use of these technologies, existing inequalities, e.g., within the workforce, are reinforced or new ones created (Huws 2014; Schrape 2021).

Many studies show that the actual use of new technologies often deviates from originally intended goals (originally Suchman 2007). Employees develop so-called 'workarounds' or use tools creatively to solve everyday work problems. Digital technologies are thus not merely 'implemented', but 'integrated' into everyday work practices. This means that the implementation in this practice should sensibly leave room for user-side appropriation and influence on implementation and further development. A participatory and context-sensitive implementation of digital technologies assumes that all those involved are able to participate in this process. This requires empowerment and an extensive recourse to the employees' labor capacity (Boes et al. 2020; Pfeiffer 2014).

This also applies to the current discussion on AI and work, where most debates on digitalization research are being continued, e.g., on questions of inequality, the automatability and replaceability of workers, qualification requirements, co-determination and design issues, new forms of work and tasks or the evolvement of new occupational fields. More recent aspects are the increased interest in ethically orientated questions, the penetration of the functional mechanisms and the question of a (re-)distribution of responsibility.

As we focus on the organizational level and issues relating to the design of AI-related governance, this does not mean leaving out the external institutions and actors that play a role in this process, such as legislation, professional and trade associations, or technology companies, but rather looking at their role in governance in organizations from the internal perspective of the organization. The concern to point out which aspects strengthen the role of employees leads us to the question of how structures of empowerment and the development of important capacities can be analyzed and assessed.

This is where structural empowerment theory comes into play. At the time, Kanter raised the question – relatively independently of technical developments – of how the structural inequality of employee groups (in this case men and women) could be offset by systematic support from the organisation and how this could ensure greater and more equal participation in developments within the organization (Kanter 1993).

Structural empowerment refers to the organizational conditions that enable employees to access essential resources, information, support, and opportunities necessary for professional autonomy and decision-making (Kanter 1993). In the context of AI integration, structural empowerment is crucial for ensuring that professionals can effectively engage with AI systems, maintain trust in their outputs, and make informed decisions. Thus, empowerment can be understood as the process of equipping employees with the necessary knowledge, tools, and institutional support to confidently interact with AI systems, critically assess their recommendations, and make informed decisions that align with ethical, professional, and organizational standards (Thomas and Velthouse 1990). Kanter's framework identifies four key dimensions of structural empowerment:

1) Access to information: information and knowledge necessary to perform tasks
2) Access to resources: assets in terms of money, material and working time
3) Access to support: reflecting own work practices by receiving guidance and feedback from colleagues and supervisors
4) Access to opportunities: learning opportunities to allow for knowledge and skills growth.

Our conceptual basis adopts these four dimensions as its guiding theoretical framework. We argue that structural empowerment is essential for employees to develop both the confidence needed to assess the trustworthiness of AI system usage and the capacity to deal with AI systems in a useful way. For example, access to information helps employees understanding AI systems' capabilities and limitations, enabling critical evaluation of outputs. Access to resources provides the necessary tools and time to engage with AI systems responsibly. Access to support through feedback from colleagues and supervisors fosters collaborative problem-solving and continuous improvement. Finally, access to opportunities ensures employees stay updated on technological advancements, adapt to emerging ethical and legal challenges and rethink the way of using AI systems.

The integration of AI systems into decision-making introduces the need for new capacities among professional knowledge workers, including physicians and managers. For instance, physicians must develop the ability to understand and explain AI-driven decisions to patients, discern when to trust AI systems' recommendations, and recognize situations where human judgment should override AI systems' output (Zhang et al. 2023).

In the medical field, the use of AI-driven diagnostic tools illustrates how the demand for professionals has shifted. For example, consider AI systems that analyze medical imaging, such as X-rays or MRIs. In the past, radiologists relied heavily on their ability to manually assess each image, leveraging their pattern recognition skills honed through years of study and experience. The process required careful attention to detail, deep expertise in imaging interpretation, and an understanding of subtle variations that indicate diseases. Today, AI systems can process and analyze large volumes of medical images rapidly, often highlighting potential issues with high accuracy (Smith and McKeen 2011). While this can reduce the time required for diagnosis, it also can shift the necessary skills for physicians. Instead of focusing solely on manual interpretation, healthcare professionals now need to: (1) be able to interpret AI-generated insights critically and validate these recommendations against their clinical judgment; (2) collaborate effectively with AI systems, which requires competencies in integrating machine-generated insights into treatment planning, ensuring seamless communication across medical teams, and maintaining adherence to clinical protocols; (3) assess potential biases in AI systems, ensure compliance with ethical standards, and prioritize patient safety.

As AI systems continue to evolve, the emphasis in healthcare shifts from manual diagnostic expertise to critical data interpretation, collaborative workflows, and ethical decision-making. But with this shift comes the need to ensure that AI systems can complement human expertise while maintaining high clinical standards and ensuring patient safety. These challenges underscore the need to empower professionals effectively to navigate this new environment – and to take part in its shaping.

Using SET as a guiding framework, the following requirements per SET dimension emerge based on the current state of our argumentation in order to enable empowerment in the context of AI systems in healthcare:

1) Access to information means that healthcare professionals and managers need clear, comprehensive, and transparent insights into how AI systems function. This includes understanding underlying algorithms, data inputs, and decision-making processes of AI to explain them effectively and make informed choices.
2) Access to resources: To work effectively with AI, healthcare professionals require resources such as state-of-the-art AI systems, adequate time to engage in AI-related training and decision-making, and institutional/organizational support for implementing AI systems.
3) For access to support continuous feedback and collaborative learning opportunities with colleagues, AI developers, and supervisors are critical. Support systems must be in place to help professionals evaluate AI-generated recommendations, troubleshoot issues, and improve their ability to work alongside AI systems.
4) Access to opportunities: Empowerment requires investment in ongoing education and training programs to help professionals and managers enhance their labor capacity. For example, workshops on interpreting AI systems' outputs, ethical usage of AI systems, and decision-making in complex scenarios can foster confidence and competence in leveraging AI systems effectively.

By structuring empowerment initiatives around these four dimensions, organizations may enable professionals to remain or become capable of acting, designing, and adapting to challenges and opportunities presented by AI systems in high-reliability decision-making environments. This approach might ensure that they can make the most of AI systems while safeguarding the integrity of their decisions, actions, and work in general.

3.2. The role of Structural Empowerment in AI Governance

As the integration of AI into healthcare goes far beyond technological change, broader questions need to be asked and implications considered and addressed. Ultimately, there are even fundamental questions of trust not only in the technology itself, but also in the organizations using it and the actors involved with their interests, knowledge, motives, scope for decision-making and action, skills and much more. And ultimately, it is also about trustworthiness at various levels and the ability to assess the trustworthiness of technical and organizational systems (at all). This concerns both patients in the healthcare sector as well as professionals from different disciplines who are (directly or indirectly) affected by the use of AI systems.

While AI systems promise enhanced efficiency, their implementation simultaneously introduces new forms of standardization and formalization, shifts in professional identity, autonomy, and (the need for) changing governance structures (Bartsch et al. 2025; Jones et al. 2023; Jungtäubl 2024). AI governance actively shapes how AI systems are embedded in clinical environments, determining who has access to knowledge, who retains decision-making authority, and – depending on the concrete governance design, directly or indirectly or intentionally or unintentionally – how professional roles change or evolve. Moreover, the tension between AI-driven neo-taylorisation of work[2] on the one hand, and increasing work intensification and rising job complexity on the other, must also be considered – particularly in healthcare settings (Altenried 2020; Mantello et al. 2023; Söllner et al. 2025).

Drawing on SET outlined before, this section explores how AI governance frameworks influence professionals' access to information, resources, support and opportunities – key dimensions relevant to their autonomy and engagement, and ultimately their ability to assess the trustworthiness of AI systems.[3] AI governance, however, is neither neutral nor purely technical. It can shape professionals' experiences with AI systems and their trust in organizational structures that (should) oversee the implementation of AI systems and not outsource it to employees choices or

2 Neo-Taylorisation of work refers to the contemporary revival and intensification of Taylorist practices – e.g., strict task decomposition, AI-driven surveillance and algorithmic control – often implemented at the expense of employee discretion and autonomy to maximize efficiency.

3 Ultimately, the employees, who in turn are in direct contact with the organizational environment – the patients – also represent trustworthiness to the outside world. They are therefore under particular pressure.

primarily to them with insufficient support.[4,5] If we now additionally extend SET through the lens of labor processes and subjectification, we can understand how AI systems influence the self-perception of professionals, their identity and the risks associated with increasing algorithmic accountability, and what impact this ultimately has on the (perceived) capacities of employees.

3.2.1. AI Governance as Structuring Force for Work, Trust(worthiness), and Empowerment

In the healthcare sector, issues relating to trust are not only about trust in technical systems such as AI reliability, but also about the implementation of embedding in and thus overarching trust in "the healthcare system", the organization of a hospital, etc. This is closely linked to governance structures that enable professionals to practically use AI in a meaningful way. "Meaningful" here implies improved productivity, enhanced work quality, healthy working conditions, self-efficacy, and professional identity. Such factors are also basic conditions for establishing and maintaining trustworthiness of AI (Kiseleva et al. 2022; Petersson et al. 2022; WHO 2021). Following SET, empowerment for 'good' AI governance[6] occurs when professionals are actively involved and have capacities to do so. Such active involvement and participation are strongly linked to SET dimensions. Professionals (and other employees as well) need transparency, explainability and contextualization in order to be able to interpret AI systems' results in a competent manner. Otherwise, there is a risk that AI governance will create a "black box bureaucracy" (Ananny and Crawford 2018; Pasquale 2015) that undermines trust and professional commitment. The active *information* of professionals is therefore essential in the context of governance and must be regulated, whereby employees should at best already be involved in the development of AI governance in order to embed their own aspiration in form of needs, knowledge and experience. It is further important to maintain and improve the *resources* of professionals by providing them with adequate training, sufficient time for professional learning and solid technical and organizational support as necessary prerequisites for enhancing their labor capacity that enables maintaining and developing AI-related competencies. AI governance structures must explicitly support these resources to reach the goal of strengthening trust, trustworthiness, professionalism, and identity. This is necessary to enable other positive effects of and for mutual learning between healthcare professionals and AI developers as well as between professionals and AI systems themselves.

---

4 According to initial, explorative empirical impressions from the DigiGov project interviews, the latter appears to be the case.

5 Ergo, questions for and perspectives on digital/AI ethics arise as discussed, e.g., Jungtäubl, Zirnig and Ruiner (2024).

6 In analogy to "good work" as a normative framing of working conditions.

Healthcare professionals need the *support* of their organization and other institutions (such as professional societies in the medical field, etc.) in the context of professional empowerment, for example in form of continuous collegial support and exchange as well as feedback mechanisms to the organization and to AI-developing companies, collaborative interactions and clearly regulated and jointly defined responsibility structures. AI governance must also prevent algorithmic fatigue (Budhwar et al. 2023) by maintaining supportive human interactions rather than replacing them with purely algorithmic controls. Ultimately, such structures and mechanisms also encourage developments around AI to promote, rather than restrict, career and professional development (*opportunities*), allowing professionals to develop new capacities through continuous, supported skill-building and co-creation processes (Jiang et al. 2021). Thus, AI governance actively shapes both the professional and organizational trust necessary for successful AI integration, enabling professionals to – finally – confidently and autonomously assess AI trustworthiness.

3.2.2. Balancing Trust, Autonomy, and Algorithmic Control

Subsequently to those considerations about effective AI governance in the medical field respectively in healthcare, three critical tensions must be carefully balanced in order to provide professional empowerment.

- *Autonomy vs. Algorithmic Standardization:* AI systems may enhance autonomy by reducing repetitive tasks; however, excessive reliance on AI systems' recommendations can undermine professional judgment and autonomy. And, autonomy can be reduced by the programming of the systems as they allow some cases and restrict others. AI governance frameworks must ensure that AI systems remain advisory rather than prescriptive or predictive, maintaining professional human agency in clinical decisions. It is also important to note that experienced staff seem less likely to use AI systems anyway and, even when they do, are good judges of how far they want to rely on AI systems' outputs, whereas less experienced staff are more likely to do so and, precisely because of their lack of experience, are less able to judge how good the quality of AI systems' outputs are. These and other points therefore have an impact on autonomy and the scope for action.
- *Competency Enhancement vs. De-skilling:* AI systems shift professional roles towards interpretative oversight. Without structured empowerment and ongoing training, professionals risk experiencing de-skilling, diminishing confidence in their own expertise and AI systems. Governance must thus emphasize capacity-based approaches, ensuring AI systems complement rather than replace professional judgment, and must address the question of what degree of dependency on those systems is acceptable.

- *Trust in AI vs. Trust in Governance:* Professionals' trust in AI systems depends significantly on transparent, accountable and participatory AI governance. Even high-performing AI systems may be rejected if AI governance lacks transparency and participatory mechanisms. AI governance structures must therefore include clear accountability processes, participatory design approaches, and explicit dispute-resolution mechanisms. Ultimately, this not only leads to a potentially improved identification with the organization and its AI governance, but increases the chance of acceptance and appropriation as well as the quality of reliance on new technology such as AI.

These tensions emerge differently depending on the complexity and apparent 'autonomy' of AI systems, which highlights the need for flexible governance approaches that are tailored to the different levels of AI integration and are continuously reviewed and realigned where necessary – from narrow to (highly) advanced AI systems.

3.2.3. Subjectification of Work, Control, and AI Governance

AI governance also directly influences professionals' subjective experiences of work, affecting self-perception, professional identity, and accountability structures. On the one hand, the introduction of AI can paradoxically lead to increased demands for autonomy (professionals interpret and integrate AI systems into work processes). On the other hand, it also harbors the risk of new forms of control and heteronomy (e.g., through algorithmic performance monitoring, prescriptive standardization). This risk sometimes arises precisely from the desired transparency, which can extend beyond the technical systems to work processes in which AI systems are embedded.

Without structural empowerment, professionals risk "algorithmic responsibilization", becoming accountable for AI systems' outcomes without decision-making authority (Adensamer et al. 2021). This leads to increased cognitive burdens, identity erosion, and challenges in overriding AI systems' recommendations – especially under economic pressures prevalent in healthcare. To counteract these negative subjectification effects, AI governance must prioritize:

- *Human-in-the-loop models:* Maintaining genuine human oversight, reducing performative validations of AI systems' outputs.
- *Co-design and participatory governance processes:* Ensuring professionals actively shape AI systems development and integration, aligning AI systems with real-world work practices.
- *Explicit accountability structures:* Clarifying responsibility for AI systems' decisions, preventing unintended accountability shifts onto frontline professionals.

4. Towards "Good" AI Governance: A Capacity-Based Approach

A genuinely empowerment-oriented AI governance model ensures that AI systems support professional autonomy, capacity-building, and trust rather than imposing rigid control or labor intensification. Key principles include:

- *Transparency and explainability:* Ensuring AI interpretability so professionals trust AI systems because they understand them, not merely due to external expectations.
- *Autonomy-preserving AI design:* AI systems as assistive technology, reinforcing clinical discretion rather than enforcing algorithmic compliance.
- *Continuous skill development and organizational learning:* Investing in continuous competence-building to develop genuine trust based on professional skill, not blind acceptance.
- *Participatory AI governance:* Involving professionals in governing AI systems actively ensures context-sensitive governance aligned with professional needs, not abstract managerial or purely technological priorities.

Embedding these principles into AI governance can create a trust-based work environment where AI reinforces rather than undermines professional expertise and autonomy.

Table 1: Structural empowerment from the perspective of AI governance.

SET Dimension: Access to...	**AI Governance Considerations**	**Impact on work, trust(worthiness), and professional identity**
Information	· Transparency regarding technical functionalities and organisational decision-making processes (*governance transparency*) · AI system explainability and interpretability of results · Clear disclosure of data provenance and decision criteria underlying AI systems' outputs	· Trust increases when professionals can autonomously assess and understand AI systems' outputs. · Enhances self-efficacy and reduces risks of "black-box bureaucracy." · Prevents risks of algorithmic responsibilization (*accountability without meaningful authority*).

SET Dimension: Access to...	**AI Governance Considerations**	**Impact on work, trust(worthiness), and professional identity**
Resources	· Investments in continuous professional development and labor capacity · Provision of sufficient time, technical infrastructure, and organisational resources · Supporting individuals in developing necessary competencies for effective AI system use and oversight	· Enables meaningful integration of AI systems into daily practices, preventing overload and de-skilling. · Builds trust as professionals possess adequate capacities and resources for safe, responsible, and efficient AI system use. · Supports healthy, sustainable, and professionally enriching working conditions.
Support	· Establishing clear accountability structures and responsibilities for AI-based decisions · Fostering collaborative and participatory AI governance models (*co-design, human-in-the-loop*) · Institutionalized support mechanisms (*peer networks, feedback systems*) to avoid algorithmic isolation and fatigue	· Reduces algorithmic isolation and cognitive burdens through collective reflection, discussion, and validation of AI recommendations. · Increases organisational trust and prevents negative subjectification effects (e.g., over-responsibilization, identity loss). · Ensures sustainable, human-centric interaction with AI systems.

SET Dimension: Access to...	AI Governance Considerations	Impact on work, trust(worthiness), and professional identity
Opportunities	· Promoting continuous development of AI competencies as integral to professional career paths and work environments · Leveraging AI systems as an enhancement (not a replacement) of professional expertise and decision-making autonomy · Creating new professional roles and opportunities through co-design processes and active professional involvement in AI system implementation	· Supports positive perception of AI systems as extending and strengthening professional competencies rather than restricting them. · Avoids negative subjectification effects (role degradation, loss of autonomy and professional discretion). · Reinforces professional identity and autonomy by enabling active shaping and meaningful integration of AI systems into professional roles.

4.1. Conceptual Framework: AI Governance, Structural Empowerment, and Professional Outcomes

The conceptual framework depicted in Figure 2 provides a structured overview of the core theoretical relationships addressed in this article. Building on previous discussions, it illustrates how organizational AI governance directly influences both organizational capacities and individual employee capacities. Central to this relationship is the SET, which serves as a critical mediator between governance structures and individual outcomes, particularly the (perceived) ability to evaluate trustworthiness of AI system usage.

At the organizational level (left side of Figure 2), AI governance structures encompass distinct dimensions such as accountability, decision rights, transparency, fairness, privacy, and data governance. These governance dimensions vary according to concrete regulations or structures within organizations as well as according to whether the AI systems are narrowly defined or more advanced and (can be) used more extensively. The interplay of these AI governance dimensions shapes the organizational conditions that professionals encounter, defining their access to crucial resources for effectively integrating AI systems into their work. The four SET dimensions provide a direct pathway through which organizational AI governance conditions influence individual capacities, shaping both practical and psychological outcomes at the workplace.

On the individual level (right side of Figure 2), the outcomes are categorized into two interconnected areas: implications for labor processes and work practices as well as subjective, psychological effects. Work practices include dimensions such as autonomy, complexity, social interactions, and meaningfulness of tasks. Individual outcomes on employees, as subjects, cover aspects like psychological ownership, role identity, perceived legitimacy of AI systems, trust, and potential negative effects such as algorithmic fatigue or feelings of over-responsibilization. Collectively, these individual outcomes determine the ability of professionals to competently assess the trustworthiness of AI systems – an essential competency increasingly demanded by regulatory frameworks such as the EU AI Act (EU, 2024). Furthermore, the conceptual framework acknowledges external influences, such as societal norms and regulations, and behavioral effects like avoidance behaviors, resistance, or deviations from prescribed AI system usage, indicating the broader social and organizational implications of AI governance decisions.

In sum, the conceptual framework clearly illustrates that AI governance is not merely a technical consideration but an essential structural dimension shaping professional roles, identities, and trust relationships within organizations. By leveraging SET, it is, on the one hand, an analyzation-tool and, on the other hand, it offers concrete pathways to actively design governance frameworks that empower professionals and foster meaningful human-AI collaboration (Chowdhury et al. 2022).

Figure 2: Conceptual Framework.

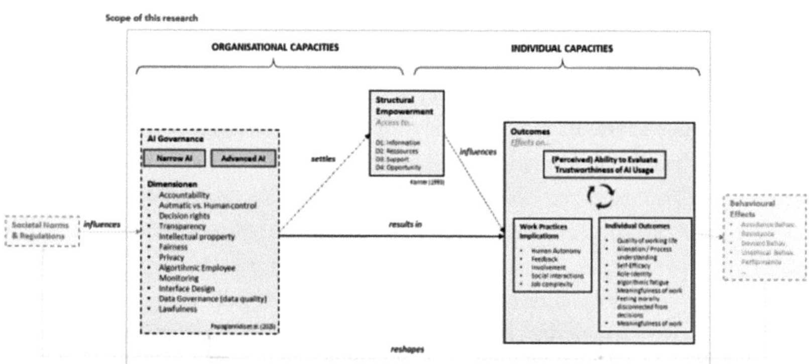

4.2. Medical AI Scenarios

Now we present two hypothetical scenarios. The first scenario focuses on a medical setting where AI-powered software assists radiologists by analyzing medical imaging, helping to identify potential health concerns like tumors. The second scenario shifts to the perspective to advanced AI systems in triage decisions. By examining these two scenarios, we will demonstrate these relationships concretely, highlighting specific governance challenges and empowerment strategies in different clinical contexts. The interplay of trust, empowerment, and subjectification varies substantially depending on the scope of the respective AI system, so the following section illustrates (some) distinctions concretely through two contrasting scenarios, demonstrating why adaptive, empowerment-oriented AI governance – grounded explicitly in SET – is essential. Such governance does more than manage technological risks; it actively shapes professional roles, work experiences, and trust in AI system integration.

4.2.1. Scenario – Narrow AI: Cancer screening and professional autonomy

Dr. Anna Berger, an experienced radiologist at a leading hospital, starts her working day assisted by an AI-powered diagnostic tool designed to enhance cancer screening accuracy. The AI system is designed to analyze MRI and CT scans, identifying potential tumors to improve diagnostic accuracy. It utilizes deep learning algorithms trained on millions of real-world medical images.

As the AI system delivers its initial analysis results, Dr. Berger notices that the AI has flagged a suspicious lesion – something she might not have noticed at first glance. She manually reviews the findings and confirms the need for further investigation. While the AI automatically detects standardized patterns in routine cases, it is up to Dr. Berger to assess complex and unusual cases, consider individual patient factors, and ultimately determine the final diagnosis and next medical steps.

This division of working tasks (and responsibilities) between human and AI illustrates how AI systems can effectively support professionals by relieving them of routine analytical tasks, thus allowing clinicians to focus on more challenging cases. However, it also increases job complexity, as Dr. Berger must now manage the intersection of her clinical judgment with the AI generated recommendations. Furthermore, the phenomenon of algorithmic opacity emerges: Dr. Berger understands that the AI system flags certain images, but she may not fully comprehend why minor variations in patient data alter AI decisions, reflecting pragmatic opacity as discussed in recent research (Bhat 2025; Shahidi Hamedani et al. 2025).

From a SET perspective, the success of such AI system integration critically depends on governance structures that ensure transparency and continuous vocational training. Dr. Berger requires targeted training, sufficient resources, and organizational support to confidently interpret, validate, and sometimes override

AI recommendations. Without clear governance guidelines, she risks experiencing heightened cognitive burdens and algorithmic responsibilization – where accountability for diagnostic outcomes is placed upon her, even as her ability to influence the underlying logic of the AI system remains limited.

Previous studies support the realism of this scenario. Eisemann et al. (2025), for instance, found that AI-assisted mammography significantly improved cancer detection rates without increasing false positives, highlighting AIs' potential to augment professional expertise without replacing it. Yet, to achieve such positive outcomes sustainably, AI governance frameworks must prioritize transparent and explainable AI, structured feedback loops, and the active participation of medical professionals in AI system development.

This scenario underscores that the perceived ability to evaluate the trustworthiness of AI systems is not merely a function of individual technical expertise, but emerges at the intersection of personal competence, institutional support, and transparent AI system design. For professionals like Dr. Berger, structural empowerment – including targeted training, resource provision, and participatory (AI) governance mechanisms – constitutes a critical precondition for developing the evaluative confidence and contextual understanding required to responsibly interpret and, where necessary, contest AI-generated outputs.

4.2.2. Scenario – Advanced AI: Automated Emergency Triage

Dr. Jonas Meier, an experienced emergency physician, begins his shift at one of the busiest emergency departments in the city. Recently, an (advanced) AI system has been assisting physicians by accessing real-time digital patient records, current hospital data, societal health statistics, and news. Its main purpose is to support triage decisions: Unlike narrow AI that operate within predefined boundaries, this AI system dynamically pulls from multiple real-time data sources to assess case urgency, prioritize patients, and allocate resources more efficiently.

The AI is capable of independently and flexibly deciding which data is relevant and which is not. It can also automatically 'decide' which tools to use in problem-solving. The AI system can independently calculate a strategy and weigh the decisions that need to be made based on that. For Jonas the AI suggestions seem to be useful, but he never knows exactly which information and tools the AI considers and which it disregards.

What sets this AI system apart is its high degree of automation: based on calculations it independently determines relevant data, analytical tools, and appropriate weighting of competing priorities. While Dr. Meier often finds its suggestions helpful, the logic behind these outputs remains opaque – raising concerns about transparency, particularly in ethically complex decisions such as prioritizing between equally urgent patients with different prognoses or resource needs.

This scenario reveals the critical importance of AI governance, particularly when advanced AI systems influence not only clinical workflows but also ethical judgment and institutional authority. In the absence of structured oversight or interdisciplinary governance bodies, clinicians like Dr. Meier are placed in vulnerable positions – legally and morally accountable, yet without real visibility into how AI outputs are calculated. This "algorithmic responsibilization" risks eroding both trust and professional autonomy. From the perspective of SET, technical reliability alone is insufficient. What is needed are institutionalized structures for explainability, clinician participation, and dynamic feedback. (AI) Governance frameworks must clearly delineate accountability lines for decision outcomes – especially when AI recommendations are accepted or overridden. Furthermore, professionals must be provided with systematic channels to shape AI system development, contribute to risk evaluation, and evaluate emerging patterns of bias or opacity.

The trustworthiness of advanced AI systems depends not only on the accuracy of these systems but also on the institutional scaffolding that ensures their contextual use. Trust arises from meaningful engagement – AI governance that ensures transparent AI integration, preserves professional discretion, and embeds human oversight as a non-negotiable standard. In this context, the perceived ability to evaluate the trustworthiness of AI (usage) becomes both more fragile and more essential. For professionals like Dr. Meier, this ability hinges not only on cognitive competencies or ethical awareness, but fundamentally on whether (AI) governance structures enable transparent insight, facilitate critical reflection, and foster a climate of shared responsibility in navigating opaque and high-reliability AI outputs.

4.3. Perceived Ability to Evaluate Trustworthiness of AI Usage

In line with the EU AI Act, professionals must not only follow AI recommendations – they must be capable of critically evaluating them. This calls for the development of user impact awareness – an emergent competence involving the capacity to foresee how AI system decisions impact patients, professional workflows, and wider organizational outcomes.

For Dr. Meier, this is particularly relevant. The advanced AI system automatically filters and interprets vast data sets but provides little insight into its 'decision rationale'. This opacity becomes problematic when prioritization outcomes are potentially biased – for instance, if patients with more complete digital histories or from wealthier neighborhoods are systematically favored. Such patterns, if undetected, can solidify through feedback loops, reinforcing disparities. Assessing trustworthiness under these conditions requires time, support, and structured empowerment:

- Explainability tools must be embedded in the AI system to reveal key decision pathways.
- Institutional mechanisms for interdisciplinary validation and ethical review are essential.

- Professionals must be granted both the time and training to engage critically with AI system outputs.

Advanced SET provides a powerful lens here: access to information (here, e.g., how the AI system works), resources (training, time), support (interdisciplinary exchange), and opportunities (participation in governance) are essential for cultivating capacity to assess AI system trustworthiness. Only through such holistic empowerment can healthcare systems ensure just, effective, and ethically grounded AI integration.

5. Conclusion

This article has examined the essential role of structural empowerment in fostering the ability to evaluate the trustworthiness of AI systems, particularly within healthcare settings. Through the lens of SET, we explored how the integration of AI systems transforms work practices, professional roles, and governance structures. Our analysis shows that while AI systems can greatly enhance decision-making efficiency, they also pose new challenges.

Successfully interacting with AI systems requires labor capacity and competencies that differ from those needed for traditional information systems. A key ability is to evaluate AI systems' trustworthiness. However, we demonstrated that this ability is not solely a matter of individual competence but also depends on organizational conditions that support vocational training, access to information, collaborative practices, and participatory governance. Using scenario-based analysis, we illustrated how narrow AI and advanced AI systems differently impact healthcare professionals' ability to make informed and responsible decisions.

Our analysis also shows that the degree of 'agency' – understood as automated and independent functioning, calculating, and 'deciding' – of the AI system significantly influences the individual capacities required to assess the trustworthiness of AI. As AI systems become more advanced and workflow automation increases, the complexity of evaluating their outputs grows, making it increasingly difficult – even impossible in some cases – for individual healthcare professionals to assess trustworthiness on their own. Therefore, AI governance should establish structures that distribute the responsibility for evaluating trustworthiness rather than placing it solely on individuals. This approach not only mitigates risks associated with increasing system autonomy but also fosters a more collaborative and supportive evaluation process.

The findings suggest that successful AI integration requires more than just technical training; it necessitates a comprehensive approach that embeds empowerment at multiple levels of the organization. Empowering healthcare professionals through transparent governance, continuous learning opportunities, and active involvement in AI governance fosters both individual and collective capacities. Thus, AI governance should not only focus on regulating technology but also on shaping the human-AI interaction in ways that reinforce professional autonomy, competence, and trust.

To achieve this, healthcare professionals need support from their organizations in the form of access to empowered structures. Practically, this means establishing new learning and design processes that enable ongoing skill development and adaptation. Digital governance should also focus on creating communication channels and reflection spaces that connect individuals and organizations, such as intranets or professional networks for continuous exchange and knowledge sharing. Designing these structures appropriately will be a core task of digital governance in the future.

Ultimately, the adoption of AI systems in healthcare must prioritize structural empowerment to ensure that professionals remain confident, capable, and accountable in their interactions with increasingly autonomous systems. This approach will enable organizations to harness the potential of AI while safeguarding human-centered decision-making and maintaining ethical standards in medical practice.

Nevertheless, our analysis shows that while structural empowerment is a promising approach to fostering skills and competencies in working with AI systems, it is not a universal solution. Creating structural conditions that ensure access to information, resources, support, and continuous development is crucial, but these measures alone cannot fully address all challenges related to AI use. In complex and dynamic environments, such as certain healthcare settings, even well-designed empowerment structures have their limitations. For example, the growing functions and functioning of advanced AI systems may continue to create uncertainties that structural measures alone cannot resolve. Moreover, continuously aligning individual competencies with rapid technological advancements remains a challenge. By embedding structural empowerment within a dynamic and adaptable governance framework, organizations can better equip healthcare professionals to navigate the evolving challenges posed by AI systems while ensuring professional autonomy, competence, and trust.

References

Adensamer, Angelika, Gsenger, Rita and Klausner, Lukas D. (2021): "'Computer says no'": Algorithmic Decision Support and Organisational Responsibility", in: Journal of Responsible Technology 7, 100014.

Altenried, Moritz (2020): "The Platform as Factory. Crowdwork and the Hidden Labour Behind Artificial Intelligence", in: Capital & Class 44(2), pp. 145–158.

Amershi, Saleema et al. (2019): "Guidelines for human-AI interaction", Proceedings of the 2019 chi conference on human factors in computing systems, pp. 1–13.

Ananny, Mike, Crawford, Kate (2018): "Seeing Without Knowing. Limitations of the Transparency Ideal and its Application to Algorithmic Accountability", in: new media & society 20(3), pp. 973–989.

Baird, Aaron and Maruping, Likoebe M. (2021): "The Next Generation of Research on IS Use. A Theoretical Framework of Delegation To and From Agentic IS Artifacts", in: MIS quarterly 45(1).

Barocas, Solon and Selbst, Andrew D. (2016): "Big data's disparate impact", in: Calif. L. Rev. 104, p. 671.

Bartsch, Sebastian et al. (2025): "Governance of High-Risk AI Systems in Healthcare and Credit Scoring", in: Business & Information Systems Engineering, pp. 1–19.

Berente, Nicholas et al. (2021): "Managing Artificial Intelligence", in: MIS quarterly 45(3).

Bhat, Maalvika (2025): "Designing AI Interfaces for Transparent Decision-Making and Ethical Reflection", Companion Proceedings of the 30th International Conference on Intelligent User Interfaces, pp. 211–214.

Bleher, Hannah and Braun, Matthias (2022): "Diffused Responsibility. Attributions of Responsibility in the Use of AI-driven Clinical Decision Support Systems", in: AI and Ethics 2(4), pp. 747–761.

Boes, Andreas et al. (2020): "Empowerment in der agilen Arbeitswelt", in: Bauer et al. (eds.), Arbeit in der digitalisierten Welt. Praxisbeispiele und Gestaltungslösungen aus dem BMBF-Förderschwerpunkt, Springer, p. 307.

Budhwar, Pawan et al. (2023): "Human Resource Management in the Age of Generative Artificial Intelligence. Perspectives and Research Directions on ChatGPT", in: Human Resource Management Journal 33(3), pp. 606–659.

Cetindamar, Dilek et al. (2022): "Explicating AI literacy of employees at digital workplaces", in: IEEE transactions on engineering management 71, pp. 810–823.

Chan, Alan et al. (2023): "Harms From Increasingly Agentic Algorithmic Systems", Proceedings of the 2023 ACM Conference on Fairness, Accountability, and Transparency, pp. 651–666.

Cheong, Ben C. (2024): "Transparency and Accountability in AI systems. Safeguarding Wellbeing in the Age of Algorithmic Decision-making", in: Frontiers in Human Dynamics 6, 1421273.

Chowdhury, Soumyadeb et al. (2022): "AI-employee Collaboration and Business Performance. Integrating Knowledge-based View, Socio-technical Systems and Organisational Socialisation Framework", in: Journal of Business Research 144, pp. 31–49.

DataRobot (2024): "The Unmet AI Needs Survey", https://www.datarobot.com/resources/the-unmet-ai-needs-survey/?utm_campaign=unmetaineedssrvyDBPR&utm_source=database&utm_medium=direct, last access: July 17, 2025.

Dietvorst, Berkeley J., Simmons, Joseph P. and Massey, Cade (2015): "Algorithm Aversion. People Erroneously Avoid Algorithms After Seeing Them Err", in: Journal of experimental psychology: General, 144(1), p. 114.

Dlugatch, Rachel, Georgieva, Antoniya and Kerasidou, Angeliki (2023): "Trustworthy Artificial Intelligence and Ethical Design. Public Perceptions of Trustworthiness of an AI-based Decision-support Tool in the Context of Intrapartum Care", in: BMC Medical Ethics 24(1), p. 42.

Donia, Joseph et al. (2024): "Lifecycles, Pipelines, and Value Chains. Toward a Focus on Events in Responsible Artificial Intelligence For Health", in: AI and Ethics, pp. 1–14.

Eisemann, Nora et al. (2025): "Nationwide Real-world Implementation of AI For Cancer Detection in Population-based Mammography Screening", in: Nature medicine, pp. 1–8.

EU (2024): "The EU Artificial Intelligence Act: Up-to-date developments and analyses of the EU AI Act", https://artificialintelligenceact.eu, last access: July 17, 2025.

Fügener, Andreas et al. (2022): "Cognitive Challenges in Human–artificial Intelligence Collaboration. Investigating the Path Toward Productive Delegation", in: Information Systems Research 33(2), pp. 678–696.

Hurley, Meghan E. et al. (2025): "Patient Consent and the Right to Notice and Explanation of AI systems Used in Health Care", in: The American Journal of Bioethics 25(3), pp. 102–114.

Huws, Ursula (2014): Labor in the Global Digital Economy. The Cybertariat Comes of Age, NYU Press.

Jiang, Lushun et al. (2021): "Opportunities and Challenges of Artificial Intelligence in the Medical Field. Current Application, Emerging Problems, and Problem-solving Strategies", in: Journal of International Medical Research 49(3), pp. 1–11.

Jones, Caroline, Thornton, James and Wyatt, Jeremy C. (2023): "Artificial Intelligence and Clinical Decision Support. Clinicians' Perspectives on Trust, Trustworthiness, and Liability", in: Medical law review 31(4), pp. 501–520.

Jungtäubl, Marc (2024): Steuerung von Arbeit durch Formalisierung [Control of Work through Formalization], Nomos.

Jungtäubl, Marc, Zirnig, Christopher and Ruiner, Caroline. (2024): "HCI Driving Alienation. Autonomy and Involvement as Blind Spots in Digital Ethics", in: AI and Ethics 4(2), pp. 617–634.

Kanter, Rosabeth M. (1993): Men and women of the corporation. New edition. Basic books.

Kapoor, Sayash et al. (2024): "Ai agents that matter", arXiv preprint arXiv:2407.01502.

Kellogg, Katherine C., Valentine, Melissa A. and Christin, Angèle (2020): "Algorithms at Work. The New Contested Terrain of Control", in: Academy of management annals 14(1), pp. 366–410.

Kiseleva, Anastasiya, Kotzinos, Dimitris and De Hert, Paul (2022): "Transparency of AI in Healthcare as a Multilayered System of Accountabilities. Between Legal Requirements and Technical Limitations", in: Frontiers in Artificial Intelligence 5, pp. 1–21.

Lebovitz, Sarah, Lifshitz-Assaf, Hila and Levina, Natalia (2022): "To Engage or Not to Engage with AI for Critical Judgments. How Professionals Deal with Opacity When Using AI for Medical Diagnosis", in: Organization science 33(1), pp. 126–148.

Mantello, Peter et al. (2023): "Bosses Without a Heart. Socio-demographic and Cross-cultural Determinants of Attitude Toward Emotional AI in the Workplace", in: AI & SOCIETY 38(1), pp. 97–119.

Matys, Thomas (2025): "Entgrenzungen und Globalisierung", in: Abels et al. (eds.), Macht, Kontrolle und Entscheidungen in Organisationen. Eine Einführung in organisationale Mikro-, Meso- und Makropolitik, Wiesbaden: Verlag für Sozialwissenschaften, pp. 165–184.Springer.

Mitchell, Margaret et al. (2025): "Fully Autonomous AI Agents Should Not be Developed", arXiv preprint arXiv:2502.02649.

Murray, Alex, Rhymer, Jen and Sirmon, David G. (2021): "Humans and Technology. Forms of Conjoined Agency in Organizations, in: Academy of management review 46(3), pp. 552–571.

Pasquale, Frank (2015): "The Black Box Society. The Secret Algorithms that Control Money and Information, Harvard University Press.

Petersson, Lena et al. (2022): "Challenges to Implementing Artificial Intelligence in Healthcare. A Qualitative Interview Study With Healthcare Leaders in Sweden", in: BMC health services research 22(1), p. 850.

Pfeiffer, Sabine (2014): "Digital Labour and the Use-value of Human Work. On the Importance of Labouring Capacity for Understanding Digital Capitalism", in: tripleC: Communication, Capitalism & Critique. Open Access Journal for a Global Sustainable Information Society 12(2), pp. 599–619.

Pinch, Trevor J. and Bijker, Wiebe E. (1984): "The Social Construction of Facts and Artefacts. Or How the Sociology of Science and the Sociology of Technology Might Benefit Each Other", in: Social studies of science 14(3), pp. 399–441.

Ramezani, Ramin et al. (2025): "Bench to Bedside. AI and Remote Patient Monitoring", in: Frontiers in Digital Health 7, pp. 1–4.

Rubeis, Giovanni (2024): "Relationships", in: Ethics of Medical AI, pp. 151–212.

Samuel, Arthur L. (1959): "Machine learning", The Technology Review 62(1), pp. 42–45.

Schrape, Jan-Felix (2021): Digitale Transformation, Bielefeld: transcript/utb.

Schuetz, Sebastian and Venkatesh, Viswanath (2020): "The Rise of Human Machines. How Cognitive Computing Systems Challenge Assumptions of User-system Interaction", in: Journal of the Association for Information Systems 21(2), pp. 460–482.

Sen, Ayusman (1999): Development as Freedom, Oxford University Press, New York.

Shahidi Hamedani, Sharareh, Aslam, Sarfraz and Shahidi Hamedani, Shervin (2025): "AI in Business Operations. Driving Urban Growth and Societal Sustainability", in: Frontiers in Artificial Intelligence 8, pp- 1–5.

Shneiderman, Ben (2020): "Bridging the Gap Between Ethics and Practice. Guidelines for Reliable, Safe, and Trustworthy Human-centered AI systems", in: ACM Transactions on Interactive Intelligent Systems (TiiS) 10(4), pp. 1–31.

Smith, Heather A. and McKeen, James D. (2011): "Enabling Collaboration with IT", in: Communications of the Association for Information Systems 28(1), p. 16.

Söllner, Matthias et al. (2025): "ChatGPT and Beyond. Exploring the Responsible Use of Generative AI in the Workplace. An Interdisciplinary Perspective", in: Business & information systems engineering, pp. 1–15.

Suchman, Lucille A. (2007): Human-machine Reconfigurations. Plans and Situated Actions, Cambridge university press.

Thomas, Kenneth W. and Velthouse, Betty A. (1990): "Cognitive Elements of Empowerment.: "An 'Interpretive' Model of Intrinsic Task Motivation", in: Academy of management review 15(4), pp. 666–681.

WHO (2021): Ethics And Governance of Artificial Intelligence For Health, https://iris.who.int/bitstream/handle/10665/341996/9789240029200-eng.pdf, last access: July 17, 2025.

Zhang, Melody et al. (2023): "The Adoption of AI in Mental Health Care–perspectives From Mental Health Professionals. Qualitative Descriptive Study", in: JMIR Formative Research 7(1), e47847.

Zou, James and Topol, Eric J. (2025): "The Rise of Agentic AI Teammates in Medicine", in: The Lancet 405(10477), p. 457.

Zuboff, Shoshana. (2023): "The Age of Surveillance Capitalism", in: Longhofer, Wesley and Winchester, Daniel (eds.), Social Theory Re-wired, Routledge pp. 203–213.

II. Governance That Enables Understanding

Right to Explanation of AI Decisions

Elena Dubovitskaya, Gregor Bosold

Abstract *As artificial intelligence (AI) systems play an increasingly significant role in decision-making across domains like credit assessment, recruitment and healthcare, concerns about their transparency and accountability are becoming more pressing. This text examines the legal notion of a right to explanation. It analyzes the recognition of this right within the framework of European Union law, particularly under the General Data Protection Regulation (GDPR), the Consumer Credit Directive, and the AI Act, and explores whether this right may also be grounded in broader fundamental rights. The analysis addresses both technical and legal strategies for tackling the challenge of AI opacity, with a particular focus on the concept of local explanations, which clarify individual decisions rather than the overall system logic. Using the example of credit scoring, the text illustrates how personalized explanations can enhance transparency and empower those affected by automated decisions.*

1. Introduction

"Nothing in life is to be feared, it is only to be understood." This famous saying, attributed to the French scientist Marie Curie, can easily be applied to the use of artificial intelligence in the 21st century. AI is now frequently used to make decisions on important and life-altering matters, such as determining credit eligibility or selecting candidates to job interviews. When a decision is negative, the affected person will often want to understand why. Only if the AI decision is adequately explained can the affected person truly comprehend it. The explanation of AI decisions is therefore an essential prerequisite for understanding their meaning. This aligns with the Capacity Approach, which emphasizes that individuals must not only have formal rights but also the actual ability to exercise them effectively.

From the legal perspective, a question arises as to whether explainability is merely a reasonable concern or a policy goal, or whether the affected individual has a legal right to know the reasons behind the decision. In practice, these reasons are often not disclosed, and legal scholarship has not paid enough attention to the right to explanation of algorithmic decisions. However, explainable and transparent AI is essential for ensuring that people can trust this increasingly powerful technology while retaining control over it.

It is therefore not surprising that, at the insistence of the European Parliament, the right to explanation was incorporated into the AI Regulation (Article 86 of the AI Act). This right has a *residual* character, meaning it applies only when no equivalent right is provided under other Union law provisions (Article 86(3) of the AI Act). The primary applicable regulations include Articles 13 et seq. of the GDPR, which establish information rights and obligations in cases of automated decision-making under Article 22(1) of the GDPR. Additionally, Article 18(8)(a) of the Consumer Credit Directive contains a specific provision for consumer credit agreements. In Germany, the legislator plans to incorporate the right to explanation in the case of scoring into the Federal Data Protection Law.

What is the scope of these provisions, and how do they align with each other? What measures should be taken to achieve an effective right to an explanation in European law to enhance the capabilities of the affected individuals? Above all, however, it is questionable whether these provisions do establish entirely distinct and independent rights to an explanation, or whether they are specific manifestations of a general principle that may derive from fundamental rights. This article seeks to explore these questions in greater depth.

2. Capacity Approach, Right to Explanation, and Explainable AI

According to the previously developed capacity-based approach, a digital system can only be considered trustworthy if people, communities or organizations are able to sufficiently understand the functioning of digital systems. However, there are limits to this understanding of digital systems: It is conceivable that the technical functioning of the systems may not be reconstructed precisely in terms of model opacity. As explained above, we can only understand systems in terms of model opacity if we are able to identify the decisive factors that led to an actual approval or rejection. The term pragmatic opacity, on the other hand, describes the additional problem that the effects and side effects of digital systems cannot be understood under the conditions of use in the social world. Explanations of AI decisions benefit multiple stakeholders, but mainly those who rely on AI for decision-making (decision-makers) and those whose rights are affected by such decisions (decision subjects). Decision-makers include, for example, physicians using AI for prognosis or corporate executives relying on AI models for business decisions. Decision subjects include consumers who receive a credit score generated by an AI-based credit agency. The right to explanation is in particular a legal tool for decision subjects, allowing them to demand local explanations from the AI system operator to overcome the model opacity.

The right to explanation for AI decisions aligns well with the Capacity Approach because it transforms a formal right into an actual capability. AI decisions are not always comprehensible to those affected. The comprehensibility depends on the complexity of the model used. If the model is simply structured (a so-called white-box model, such as linear regression or a simple decision tree), its decisions are easily understandable for users. Large and powerful AI models, such as artificial neural networks or large, complex decision trees, are generally considered black boxes (Dubovitskaya and Buchholz 2023: 64–65). With increasing performance comes a loss of transparency. The output of such black-box models is not comprehensible for users, whether they are IT professionals, domain experts, or laypersons.

Without clear and understandable explanations, individuals affected by automated decisions may lack the knowledge or means to challenge unfair outcomes, thereby limiting their actual agency. Ensuring real access to justice and redress means that individuals can understand why a decision was made about them and act if necessary. Without this right, a person denied a loan due to an algorithmic credit score would have no real way to challenge or improve their situation. The right to explanation also promotes transparency and reduces power imbalances. Many algorithmic decisions are made by companies or institutions that hold significantly more knowledge and influence than the individuals affected. Providing explanations helps to reduce asymmetries of information and allows people to participate in decisions that impact their lives.

Another important aspect is supporting informed decision-making. If individuals understand the factors affecting their credit score or job application, they can make better choices about how to improve their situation. Additionally, protecting vulnerable groups is a crucial concern, as marginalized individuals, such as those with lower digital literacy or from disadvantaged socioeconomic backgrounds, are particularly at risk of being unfairly affected by algorithmic decisions. The right to an explanation ensures that such individuals are not left without recourse, thereby supporting a fairer distribution of capabilities.

However, the right to an explanation can only support affected individuals and enhance their capabilities if it is technically feasible. From a technical perspective, the problem of model opacity can generally be addressed in various ways. One approach to solving this problem is to seek global ex-ante explanations of the model's behaviour, which could make its decisions predictable. According to the current state of technology, this is largely infeasible, a point frequently emphasized in legal literature (Kumkar and Roth-Isigkeit 2020: 285). As a result, some propose significantly lowering the requirements for explainability, arguing that a rudimentary understanding of the basic functionality of an AI model suffices to establish "algorithmic control" (Weber, Kiefner and Jobst 2018: 1132). This perspective is found in both corporate and data protection law scholarship. Others counter that such a superficial understanding is ineffective and advocate for the abandonment of explainability requirements altogether (Wischmeyer 2018: 53).

Nevertheless, the problem of model opacity can be addressed not only globally but also locally. The latter approach focuses on explanations that do not predict future model decisions but instead clarify decisions already made (local explanations). This approach does not require reconstructing the technical workings of the model but merely ensures that individual decisions can be understood. Various technical methods for explainable AI (XAI) are already available for this purpose. XAI tools are generally software solutions, specifically developed to provide explanations for the decisions of AI models. These tools consist of algorithms that can either be integrated into existing AI models or used separately to interpret model outputs. Most of these tools are available as libraries or frameworks that data scientists and AI developers can incorporate into their workflows. These methods include, for example, LIME (Local Interpretable Model-Agnostic Explanations, see Ribeiro, Singh and Guestrin 2016) and SHAP (SHapley Additive exPlanations, Lundberg and Lee 2017). A major advantage of both methods is that they are model-agnostic, as indicated by the term "Model-Agnostic" in LIME. This means they can be used to explain any AI model, regardless of its structure. Additionally, they are capable of handling various data formats, including image, text, and tabular data. While LIME is computationally faster and easier to implement for quick local explanations, SHAP offers a more theoretically justified, consistent, and robust approach to AI explainability. SHAP is mathematically well-founded: The method is based on concepts from game theory, particularly the Shapley values, which were developed in the 1950s by mathematician Lloyd Shapley. These values provide a fair method for distributing gains (or contributions) among participants in a cooperative game. In practical terms, SHAP works by calculating the contribution of each feature, such as gender or age, to a specific model prediction. This is done by measuring the difference in the prediction when a particular feature is omitted or included. A positive SHAP value indicates that the given feature increases the prediction, while a negative value suggests that it decreases it. The magnitude of the SHAP value reflects the strength of the feature's influence. This approach determines how a particular feature affects a specific model prediction, a process known as feature attribution. SHAP also considers interactions between different features, revealing dependencies where the contribution of one feature depends on whether other features are present or not.

Nonetheless, it is imperative to acknowledge that this XAI method also has weaknesses. For complex data, the computation can take a long time because the number of possible feature combinations increases exponentially with the number of features in the dataset. This issue has been partially mitigated with methods like Kernel-SHAP, which approximates Shapley values and thus speeds up the computation. Moreover, while long computation times can be problematic for time-critical real-time applications, they are less of a concern when explaining decisions that have already been made.

A more fundamental problem, however, is that SHAP, like nearly all currently available XAI methods, does not guarantee that the generated explanations accurately reflect the behaviour of the AI model being explained. One possible solution could be to combine different methods, such as SHAP with counterfactual explanations. Counterfactual explanations (see, for example, in the context of GDPR Wachter, Mittelstadt and Russell 2017: 841 et seq.) provide insights into AI decisions by answering "what if" questions. Instead of describing why a particular decision was made, they explain how the decision could have been different. A counterfactual explanation identifies the smallest possible change in input data that would have led to a different outcome. For example, in a loan approval scenario, if an applicant is denied a loan, a counterfactual explanation might state: "If your annual income had been € 5,000 higher, your loan would have been approved."

If both methods produce consistent results, the likelihood that the explanations are accurate increases. Additionally, combining both methods can improve the quality of explanations: SHAP can provide an overview of feature importance, while counterfactual explanations offer actionable insights for individuals affected by AI decisions. For example, in credit scoring, SHAP could explain that income is the most important factor, while a counterfactual explanation could specify exactly how much income needs to increase for loan approval.

Nonetheless, it should be noted that even if multiple XAI methods produce consistent results, they may still deviate from the actual decision-making process of the AI model. However, this residual uncertainty seems acceptable, as there is no guarantee of absolute correctness in human explanations either. When a person justifies their decision, we cannot know to what extent their explanation truly reflects their internal decision-making process. Even if the explanation is given honestly, it may merely be an attempt to retrospectively rationalize, what was actually the result of intuitive and instinctive decision-making.

3. Right to an Explanation in European Secondary Law

After demonstrating that the right to an explanation can, in principle, be technically implemented, it is worthwhile to examine its specific manifestations in European secondary law, such as EU Regulations and Directives, more closely.

3.1. Right to Explanation under the GDPR

3.1.1. Protection of individuals affected by automated decisions

The GDPR provides a special right of access for data subjects in cases where decisions "based solely on automated processing" have legal effects on them or significantly affect them in a similar manner (Article 22(1) of the GDPR). This right is established in Article 15(1)(h) of the GDPR, which requires that affected individuals receive "meaningful information about the logic involved, as well as the significance and the envisaged consequences of such processing for the data subject". The corresponding obligations for data controllers are set out in Article 13(2)(f) of the GDPR (for data collected directly from the data subject) and Article 14(2)(g) of the GDPR (for data collected from third parties).

The scope of the right under Article 15(1)(h) of the GDPR has always been controversial. The main point of uncertainty was the interpretation of the term "the logic involved". Some scholars took the expression literally and sought to highlight the technical challenges of obtaining meaningful information about the logic of certain machine learning algorithms. Consequently, they concluded that "the game is not worth the candle" and that "a 'right to an explanation' is probably not the remedy you are looking for" (Edwards and Veale 2017: 65). Others went even further and argued that a right to explanation of automated decision-making does not exist in the GDPR (Wachter, Mittelstadt and Floridi 2017: 90). In line with this, some data protection scholars in Germany opposed an extensive right of access, arguing that a general description of the data processing was sufficient (Kumkar and Roth-Isigkeit 2020: 277; Kamlah 2023: 28; Bienemann, 2022: 38). Others countered that data subjects have a right to an explanation and should also be informed of the weighting of different factors that led to the decision, the identity of reference groups, and the reasons for their categorization within such groups (Buchner 2024: 35a; Dix 2019: 25).

The European Court of Justice (CJEU) has addressed this issue in the context of credit scoring, which credit agencies perform on behalf of other companies, such as banks or mobile network providers. Credit agencies often use mathematical methods like logistic regression to predict the influence of certain variables, such as income, the number of loans, or changes of residence, on a person's creditworthiness. Strictly speaking, such methods do not fall under AI, nor do they involve machine learning algorithms. However, since the individual score is calculated automatically, this constitutes profiling within the meaning of Article 4(4) of the GDPR, which refers to a form of automated processing of personal data consisting of the use of personal data to evaluate certain personal aspects relating to a natural person, in particular to analyse or predict aspects concerning that person's creditworthiness.

In 2023, the CJEU ruled that the credit score calculated by Schufa Holding, Germanys largest credit agency, constitutes an automated decision under Article 22(1) of the GDPR (CJEU 7.12.2023 – *SCHUFA Holding (Scoring)*, para 50). With this ruling, the CJEU rejected the prevailing view in Germany, which, in line with the wording of Article 22(1) of the GDPR, distinguishes between a "decision" on the one hand and the "profiling" on the other. According to this view, profiling, and thus credit scoring, is generally not considered a decision but merely a preceding data analysis process (profiling) that serves as a basis for the decision (BGH 28.1.2014, para. 34; Schulz 2022: 4, 17; Spindler and Horváth 2019: 12). Implicitly, the CJEU appears to have decided to distinguish between profiling (scoring) and its outcome (the score), interpreting the latter as a "decision". According to the CJEU, this decision significantly affects the data subject within the meaning of Article 22(1) of the GDPR. It applies at least when the score is low, as companies that have commissioned SCHUFA to assess creditworthiness, in practice, almost always decide against entering into a contract with the data subject in such cases.

Following this decision, credit agencies are now required to provide explanations under Article 15(1)(h) of the GDPR (Dubovitskaya and Bosold 2024: 1809). In response to these extensive changes, SCHUFA has announced a transparency initiative aimed at revising its existing scoring practices (Schufa 2024).

Another significant ruling by the CJEU followed in February 2025 (CJEU, 27.2.2025 – *Dun & Bradstreet Austria*). The case was based on a dispute between an Austrian consumer and the credit reporting agency Dun & Bradstreet Austria. The consumer, whose creditworthiness was rated poorly by Dun & Bradstreet, claimed that her right of access under Article 15(1)(h) of the GDPR had been violated. In its decision, the CJEU provided a detailed interpretation of the scope of this right of access and addressed the question of how a balance should be struck between this right on the one hand and the interests of data processors, particularly the protection of their trade secrets, on the other.

The CJEU has decided that, in order to provide the "meaningful information about the logic involved", the controller must explain to the data subject, in a concise, transparent, intelligible and easily accessible manner, the procedure and principles actually applied in the automated processing of his or her personal data in order to obtain a particular result – for example, a credit profile. The core element of the explanation under Article 15(1)(h) of the GDPR is therefore "the procedure and principles" applied in the automated data processing (CJEU 27.2.2025 – *Dun & Bradstreet Austria*, Ruling – Point 1). Advocate General de la Tour previously referred to "the essential elements of the method and the criteria used" (Advocate General de la Tour 12.9.2024 – *Dun & Bradstreet Austria*, para. 64).

The CJEU does not explicitly define what is meant by "the procedure and principles". However, it establishes a clear link between Article 15(1)(h) of the GDPR and the rights of the data subject under Article 22(3) of the GDPR. The right of access under Article 15(1)(h) of the GDPR primarily serves to enable the data subject to effectively exercise their rights under Article 22(3) of the GDPR, namely the right to express their own point of view and the right to challenge the decision (CJEU, 27.2.2025 – *Dun & Bradstreet Austria*, para. 55). In order to effectively exercise these rights, the data subject must be able to understand the reasons for the automated decision (CJEU, 27.2.2025 – *Dun & Bradstreet Austria*, para. 56). From this, the CJEU, for the first time in the history of the GDPR, derives a genuine right to an explanation: "[...] Article 15(1)(h) of the GDPR affords the data subject a genuine right to an explanation as to the functioning of the mechanism involved in automated decision-making of which that person was the subject and of the result of that decision" (CJEU, 27.2.2025 – *Dun & Bradstreet Austria*, para. 57).

In contrast, it is not necessary to provide the data subject with a complex mathematical formula (such as an algorithm) or to describe every step of an automated decision-making process in detail. Such information does not meet the requirements for explanations under Article 15(1)(h) of the GDPR, as it would not constitute a sufficiently precise and comprehensible clarification. Instead, the controller must describe the procedure and principles actually applied in a way that "the data subject can understand which of his or her personal data have been used in the automated decision-making at issue, with the complexity of the operations to be carried out in the context of automated decision-making not being capable of relieving the controller of the duty to provide an explanation" (CJEU, 27.2.2025 – *Dun & Bradstreet Austria*, para. 59, 61).

These findings of the Court provide valuable guidance for the right to an explanation and its technical implementation. First, the CJEU establishes that in the case of automated decisions, including AI decisions, there is a "genuine right to an explanation". Second, this right does not relate to the general functioning of the AI model used, but rather to the specific decisions made by the model (local explanations). Third, the CJEU does not prescribe a specific XAI method for explanations, nor does it formulate concrete requirements that would indicate the necessity of a particular method, such as SHAP, for example. Instead, the Court limits itself to the requirement that the decision must be comprehensible to the data subject.

However, in its judgment, the Court states that in a case of profiling, as in this instance, it may be sufficiently transparent and intelligible to inform the data subject of the extent to which a variation in the personal data considered would have led to a different result (CJEU, 27.2.2025 – *Dun & Bradstreet Austria*, para. 62). This means, from a technical perspective, that counterfactual explanations may also be sufficient. In the Schufa case, however, Advocate General Pikamäe took the view that the controller must inform the data subject about "the factors taken into account for the decision-making process and their respective weight on an aggregate level" (Advocate General Pikamäe 16.3.2023 – *SCHUFA Holding (Scoring)*, para. 58). The mention of the "factors taken into account" (= features) and their weighting (= contribution to a specific model prediction) suggested that methods such as SHAP must be used to explain ML decisions, as they reveal feature importance. The CJEU does not appear to require this.

The requirement that the explanation has to be provided in a concise, transparent, intelligible, and easily accessible form aligns well with the Capacity Approach, which emphasizes that individuals must not only have formal rights but also the real ability to exercise them effectively. Merely providing information is insufficient if people struggle to comprehend it. Therefore, the CJEU's decision bridges the gap between formal rights and real capabilities, making the right to an explanation practically accessible and meaningful.

3.1.2. Protection of trade secrets

The other major part of the CJEU judgement in Dun & Bradstreet case addresses the conflict between the right to an explanation and the protection of trade secrets. The CJEU emphasizes that the right of access must not infringe on the rights and freedoms of others, such as trade secrets or intellectual property rights, including copyright on software. On the other hand, these rights cannot be used as a justification to completely deny the data subject any access to information (CJEU, 27.2.2025 – *Dun & Bradstreet Austria*, para. 69, 70).

For this reason, the CJEU seeks to establish a fair balance. In doing so, it adopts the solution advanced by the referring Austrian court in its request for a preliminary ruling, as subsequently supported by Advocate General de la Tour in his Opinion. If the controller considers that the information required to be provided to the data subject under Article 15(1)(h) of the GDPR constitutes a trade secret, they must submit this information to the competent supervisory authority or court, which must then weigh the competing rights and interests to determine the extent of the data subject's right of access under Article 15 of the GDPR (CJEU, 27.2.2025 – *Dun & Bradstreet Austria*, Ruling – Point 2). Especially when the court or the competent authority concludes that the information is protected as a trade secret, a decision must be made based on the principle of proportionality regarding which information, and to what extent, must be disclosed to the data subject (see CJEU, 27.2.2025 – *Dun & Bradstreet Austria*, para. 73).

The recognition that fundamental rights must be balanced with conflicting rights on the other side is not new in the CJEU's case law (CJEU 29.1.2008 – *Promusicae*, para. 65 et seq.; CJEU 29.7.2019 – *Spiegel Online*, para. 51 et seq.; CJEU 4.5.2023 – *Austrian Data Protection Authority*, para. 44). It is only questionable whether the CJEU achieves a fair and practical balance in Dun & Bradstreet. The CJEU is correct in stating that balancing these conflicting rights may require limiting explanation obligations to the minimum information necessary to protect affected individuals' rights. However, since this does not require disclosing the actual algorithm or source code (see above), direct exposure of trade secrets is unlikely. Only indirect exposure through reverse engineering based on explanations could pose a risk.

Against this background, it is a positive development that the burden of proof for the risk of trade secret disclosure now lies with the controller, not the affected individual. The controller can no longer simply claim that explaining the automated decision would reveal trade secrets as a blanket defence. Instead, they must convince the court or the competent authority that the information in question is indeed protected. On the other hand, the court or the competent authority will often lack the technical expertise to thoroughly assess the controller's claims (Metikoš and Ausloos 2025: 26; Langenbucher and Bauer 2025). This is particularly true for the question of whether and when the reconstruction of the original algorithm through reverse engineering is possible. This will lead to the involvement of expert opinions, which, in turn, could result in delays and legal uncertainty.

3.2. Right to Explanation in the Consumer Credit Directive

The Consumer Credit Directive stipulates that the creditor must be required under national law to carefully assess the consumer's creditworthiness before concluding a credit agreement (Article 18(1) of the Consumer Credit Directive). The assessment of creditworthiness shall be carried out on the basis of relevant and accurate information regarding the consumer's income, expenses, and other financial and economic circumstances, which must be necessary and proportionate to the nature, duration, value, and risks of the credit for the consumer. The information shall be obtained from relevant internal or external sources (Article 18(3) of the Consumer Credit Directive).

Automated processing of personal data (internal or external scoring) is also permitted in this context. However, in such cases, the consumer has the right under Article 18(8)(a) of the Consumer Credit Directive to "request and obtain from the creditor a clear and comprehensible explanation of the assessment of creditworthiness, including the logic and risks involved in the automated processing of personal data, as well as its significance and effects on the decision." This provision constitutes a specific regulation for consumer credit agreements, which takes precedence over the right of access under Article 15(1)(h) of the GDPR (see also Buck-Heeb 2023: para. 41).

Since the wording of Article 18(8)(a) of the Consumer Credit Directive is similar to that used in Article 15(1)(h) of the GDPR, the scope of this right of access can be interpreted in light of the CJEU's decision in the Dun & Bradstreet Austria case (see above). This means that the consumer must be provided with an explanation of the automated decision regarding their creditworthiness in a way that allows them to understand it.

It should be noted that Article 18(8)(a) of the Consumer Credit Directive covers a broader range of automated decisions than the GDPR. The GDPR requires a decision to be based "solely" on automated processing (see Article 22(1) of the GDPR, as referenced by Articles 13(2)(f), 14(2)(g), and 15(1)(h) of the GDPR). Recital 71 of the GDPR refers in this context to decisions made "without any human intervention". However, decisions in which human involvement is merely formal are also included, such as cases where a person processes a decision but has no ability to deviate from an automatically generated outcome (Buchner 2024: 15). It should also be noted that, according to the Schufa judgement of the CJEU, the results of fully automated profiling could also constitute "decisions" within the meaning of Article 22(1) of the GDPR (see above). This significantly expanded the scope of the provision.

Under the Consumer Credit Directive, it is sufficient that the creditworthiness assessment "involves" the use of automated processing of personal data (Article 18(8) of the Consumer Credit Directive). This means that the right to an explanation of the creditworthiness assessment applies even if automated data processing is merely a component of the assessment.

3.3. Right to Explanation in the AI Act

For certain high-risk AI systems, the right to an explanation is also provided for in Article 86(1) of the AI Act. However, this right to explanation has a supplementary character: It applies only insofar as no equivalent right is provided under other provisions of Union law (Article 86(3) of the AI Act).

Article 86(1) of the AI Act applies to so-called high-risk AI systems listed in Annex III of the AI Act, with the exception of AI systems intended to be used as safety components in the management and operation of critical digital infrastructure, road traffic, or the supply of water, gas, heating, or electricity. In contrast, the high-risk AI systems covered include those in the areas of biometrics, education and vocational training, employment, workers' management and access to self-employment, essential services (such as healthcare, evaluation of creditworthiness, life and health insurance, emergency calls and services), law enforcement, migration, asylum and border control management, as well as the administration of justice and democratic processes.

If a decision is taken based on the output of such a high-risk AI system, any affected person subject to the decision has the right to obtain from the deployer clear and meaningful explanations of the role of the AI system in the decision-making procedure and the main elements of the decision taken. The prerequisite is that the decision produces legal effects or similarly significantly affects the person in a way that they consider to have an adverse impact on their health, safety, or fundamental rights.

The obligation to provide an explanation applies to deployers, meaning persons, public authorities, agencies, or other bodies using an AI system under their authority, except where the AI system is used in the course of a personal, non-professional activity (Article 3(4) of the AI Act). The right to an explanation is available to every affected person. Despite the broad wording ("person"), it is generally understood that only natural persons are covered by this provision (Hartmann 2024: 12).

Article 86(1) of the AI Act applies to both fully and partially automated decision-making processes, as it refers to decisions made by the deployer "on the basis" of the system's output. The scope of the right to an explanation is therefore broader than under the GDPR and similar to that in the Consumer Credit Directive. Similar to Article 22(1) of the GDPR, Article 86(1) of the AI Act requires that the decision has legal effects on a person or similarly significantly affects them. Given the nature of high-risk AI systems to which this provision applies, this will regularly be the case. These are AI systems used in highly sensitive areas.

Additionally, from the perspective of the affected individual, the impact must relate to health, safety, or fundamental rights. The range of potentially affected fundamental rights is broad. It includes human dignity, respect for private and family life, protection of personal data, freedom of expression and information, freedom of assembly and association, the right to non-discrimination, the right to education, consumer protection, workers' rights, the rights of persons with disabilities, gender equality, intellectual property rights, the right to an effective remedy and a fair trial, the right to a defence, the presumption of innocence, and the right to good administration (see Recital 48 of the AI Act). According to the wording of Article 86(1) of the AI Act, it is sufficient that the affected person perceives the impact to be adverse. Therefore, it should be enough for the individual to credibly demonstrate the impairment (Hartmann 2024: 13).

As indicated by Article 86(1) of the AI Act, the affected person has the right to receive a clear and meaningful explanation from the operator regarding the role of the AI system in the decision-making process and "the main elements" of the decision made. However, the regulation does not specify what exactly is meant by "the main elements" (critically also Merkle 2024: 42). There are good reasons to interpret this as referring to the most important factors (Hartmann 2024: 13), including their weighting. It remains to be seen whether the CJEU will extend its interpretation of Article 15(1)(h) of the GDPR, as formulated in Dun & Bradstreet, to Article 86(1) of the AI Act.

4. Statutory Right to Explanation in German Law

The German legislator intends to take the new European developments into account when reforming the Federal Data Protection Act (FDPA). Increasing transparency in credit scoring had already been planned in Germany for some time (Coalition Agreement 2021–2025: 17). The regulation of credit scoring is outlined in the new § 37a of the FDPA, which is designed as an exception to the prohibition under Article 22(1) of the GDPR (Parliament 2024: 22). It should be remembered that Article 22(1) of the GDPR generally prohibits decisions based solely on automated data processing that produce legal effects concerning the data subject or similarly significantly affect them. Article 22(2)(b) of the GDPR provides an exception to this prohibition if the decision is authorized by Union or Member State law to which the controller is subject, and which also lays down suitable measures to safeguard the data subject's rights, freedoms, and legitimate interests. The new § 37a of the FDPA will provide such authorization.

As safeguards for the data subject's rights, freedoms, and legitimate interests, § 37a of the FDPA (draft version) introduces new substantive and procedural requirements for scoring. In particular, § 37a(4) of the FDPA (draft version) regulates the disclosure obligations of controllers in the context of scoring, given that scoring involves "special risks for the data subject, the impact and scope of which often cannot be understood without additional information" (Parliament 2024: 24). The disclosure obligations cover four categories of data: (1) the personal data and criteria used to generate the probability values; (2) the weighting of categories and criteria, as well as the relationships between individual criteria, that have the greatest influence on the probability value; (3) the significance of the specific probability value; (4) the generated probability values and their recipients.

The explanatory notes to the law provide further clarification only with regard to the third category, the significance of the specific probability value. To make this understandable, specifying a Gini coefficient may be considered; additionally, the probability value should be compared to reference values from other segments of the population (Parliament 2024: 24).

According to the proposed § 37a(4) of the FDPA, the information must be made available to the data subject upon request and "in a precise, transparent, understandable, and easily accessible form in clear and simple language", if necessary, in a language tailored to the target group. Just like similar statements made by the CJEU in the Dun & Bradstreet case (see above), this can be seen as a positive aspect from the perspective of the Capacity Approach. Ensuring that information about credit scoring is communicated in a way that different groups can understand strengthens individuals' capability to make informed financial decisions. A target-group-specific approach takes into account varying levels of financial literacy and language proficiency, helping to reduce structural inequalities and ensuring that transparency measures lead to actual empowerment.

The right to explanation under the proposed § 37a(4) of the FDPA generally exists alongside the right to explanation under Article 15(1)(h) of the GDPR. While this duplication presents certain legal and practical disadvantages, the creation of a clearer and more detailed regulation at the national level is nevertheless welcome, given the ongoing uncertainties in the interpretation of the European provision. At least in the area of scoring, the proposed § 37a(4) sentence 1 no. 2 of the FDPA clarifies that the weighting of data that most significantly influences the probability value (feature importance) must be disclosed. Additionally, to explain the significance of an individual score, it must be placed in relation to reference values from other segments of the population, which implies the disclosure of comparison groups.

The exclusion of the right of access provided for in the new § 34(1) of the FDPA to protect trade secrets shall not apply to scoring, as stated in § 37a(5) of the FDPA (draft version). The legislative materials justify this by referring to the particular risks to which the affected individuals are exposed in the context of scoring (Parliament 2024: 24). In light of the CJEU's decision in the Dun & Bradstreet case, completely eliminating the protection of trade secrets in the context of scoring goes too far. Such a provision would disregard the fundamental rights of scoring providers, which are enshrined in the EU Charter of Fundamental Rights. Consequently, it would not be compatible with higher-ranking European law.

5. Fundamental Rights Foundations of the Right to Explanation

Given the various statutory manifestations of the right to explanation, the question arises as to whether these different forms share a common origin in the fundamental rights of the affected individual. In this respect, it is worth taking a closer look at the Charter of Fundamental Rights of the European Union (CFR) and the European Convention on Human Rights (ECHR).

Article 8 of the CFR guarantees the right to the protection of personal data. The connection between Article 8 CFR and the GDPR is regularly emphasized by the CJEU. The aim of the GDPR is to ensure a high level of protection for the right to privacy in the processing of personal data (CJEU 4.10.2024 – *Schrems*, para. 45). The right to the protection of personal data includes the principle of transparency, ensuring that individuals are informed about the processing of their data and have control over it. The right to an explanation emerges as a prerequisite for effectively exercising this control. The issue of control is particularly pressing when data processing is carried out by AI, especially through machine learning algorithms.

There are several reasons for this. In human understanding, people should control machines, not the other way around. The reversal of this dynamic evokes fear, a theme explored in numerous science fiction novels and films (Seng 2018: 60 et seq.). The fear of being reduced to an object by a machine can only be mitigated by a sense of control, which transforms us into active subjects and restores our human dignity. The comprehensibility of machine decisions helps to regain real control, rather than merely the illusion of control (Dubovitskaya and Buchholz 2023: 71).

Another reason why the risk of losing control is particularly high lies in the specific characteristics of machine learning algorithms. Their decisions are not causal conclusions, but rather probabilistic calculations based on the analysis of vast amounts of data. As a result, machine learning applications can make errors that a human would not. These errors may arise from poor data quality, faulty training, or even from the system becoming the target of a cyberattack – a risk that should not be underestimated. Further, there is a risk of discrimination due to biased algorithms or inaccurate data. These particular risks can be mitigated by the right to an explanation, as it can reveal hidden deficiencies in decision-making and thus protect affected individuals from the risks described. In particular, the right to an explanation can help uncover and challenge discriminatory patterns in decision-making. In this way, the right to an explanation also contributes to enforcing the prohibition of discrimination, which is enshrined in Article 21 of the CFR.

It is worth noting that the German Federal Constitutional Court (FCC) already addressed the issue of loss of control in connection with the novel possibilities of data processing in its decision "Right to Be Forgotten I" (FCC 6.11.2019 – *Right to be forgotten I*, para. 90). The FCC considers it the task of fundamental rights, to protect individuals from such a loss of control. In this context, the court refers to the general right of personality, which is derived from Article 2(1) and Article 1(1) of the German Constitution. The general right of personality protects, among other things, informational self-determination and thus represents a national counterpart to the right to the protection of personal data under Article 8 of the CFR.

According to the "Right to Be Forgotten I" ruling, the right to informational self-determination is primarily to be understood as a guarantee that protects individuals against non-transparent data processing and usage. The court states: "It provides protection against third parties seizing individual data and using it in an incomprehensible manner as an instrument to categorize affected individuals into characteristics, types, or profiles over which they have no control, but which are of significant importance for the free development of personality and equal participation in society" (FCC 6.11.2019 – *Right to be forgotten I*, para. 90).

In the cited part of the "Right to Be Forgotten I" ruling, the FCC refers to profiling. It is important to note that profiling is not entirely identical to AI-based decision-making: AI decisions can be made with or without profiling, and profiling can occur without AI. However, at the very least, when an AI decision involves profiling – which is often the case in practice – the FCC likely takes the position that algorithmic calculations must be made comprehensible. This position is also valuable from the perspective of the Capacity Approach. By explicitly protecting individuals against opaque data processing practices that can categorize them without their control or understanding, it directly supports individuals' capability to actively shape their own lives and participate equally in society. Ensuring transparency and comprehensibility in data-driven decisions empowers individuals with genuine agency and reduces structural inequalities stemming from asymmetric information and power dynamics.

Furthermore, there is a close connection between the right to an explanation and the fundamental right to an effective remedy and to a fair trial (Article 47 of the CFR).

In order for an affected person to exercise their right to an effective remedy, they must be able to understand the basis on which the decision was made. The right to an explanation thus serves as a key instrument in upholding the right to be heard. In an earlier decision, the CJEU recognized the tension between the right to effective remedy and the opacity of conventional AI technologies. The lack of transparency in AI decision-making systems, the Court noted, "could [...] deprive the affected individuals of their right to an effective judicial remedy enshrined in Article 47 of the CFR" (CJEU 21.6.2022 – *Ligue des droits humains*, para. 195; Bäcker 2024: 27c). In Dun & Bradstreet, the CJEU also refers to Article 47 of the CFR when balancing disclosure and the protection of trade secrets, stating that disclosure is necessary to ensure the effective exercise of the right to an effective remedy (CJEU, 27.2.2025 – *Dun & Bradstreet Austria*, para. 73).

In conclusion, it must be recognized that the right to an explanation is not merely guaranteed by ordinary (European and national) laws but has constitutional roots. However, it would go too far to regard it as an independent fundamental right. Rather, it is an expression of various rights, such as the right to the protection of personal data and the right to an effective remedy.

6. Conclusion, Critique, and Outlook

Europe has made significant progress in establishing the right to explanation of AI Decisions. This development is crucial for fostering trustworthy AI "made in Europe." The right to an explanation also plays a central role within the Capacity Approach. The Capacity Approach emphasizes that people should not only have formal rights but must also be genuinely able to exercise them effectively. The right to an explanation serves this very purpose by ensuring that affected individuals can understand and, if necessary, challenge automated decisions.

However, the current state is not yet optimal. Although the right to explanation is now recognized in various EU legal acts, such as the GDPR, the Consumer Credit Directive, and the AI Act, as well as in the case law of the CJEU, its current design remains inadequate and problematic. The existing regulations leave several critical questions unanswered, particularly regarding their coherence, practical implementation, and enforceability.

First, there is still no precise definition of what constitutes an "explanation." In the Dun & Bradstreet ruling, the CJEU refers to "the procedure and principles" applied in automated data processing without defining what these "procedure and principles" actually are. The court merely clarifies that the decision must be made comprehensible to the affected person without formulating concrete requirements for explanations. This could incentivize companies to provide formal but largely uninformative explanations.

Second, the tension between transparency and the protection of trade secrets remains partially unresolved. The CJEU delegates this issue to national courts and supervisory authorities, which, due to a lack of technical expertise, may not be able to assess whether the controller's claim that explaining automated decisions would lead to the disclosure of trade secrets is valid. This could result in courts and supervisory authorities increasingly relying on expert opinions, leading to delays in proceedings and legal uncertainty. Additionally, it is likely that different judicial and administrative practices will emerge across Member States, further complicating legal consistency on this issue.

Third, there is an inconsistency between different EU laws. While the right to an explanation under Article 15(1)(h) of the GDPR requires a decision to be based "solely" on automated processing, the Consumer Credit Directive and the AI Act consider it sufficient if the decision is made with the use of automated systems. As a result, the level of protection for affected individuals varies depending on the applicable legal framework. On the other hand, the Consumer Credit Directive and the AI Act are narrower in scope than the GDPR. Article 18(8)(a) of the Consumer Credit Directive applies only to creditworthiness assessments in consumer credit agreements. Similarly, Article 86(1) of the AI Act applies only to certain high-risk AI systems. In contrast, the right to an explanation under Article 15(1)(h) of the GDPR generally applies to any automated decision, meaning that, in theory, an affected person could even request an explanation for personalized advertising. The necessary limitation is then established through the criterion of legal effect or a similarly significant effect (Article 22(1) of the GDPR). It remains to be seen whether this is sufficient to prevent the abusive exercise of the right to explanation under the GDPR.

In the AI Act, the European legislator missed the opportunity to establish clear regulatory requirements for explaining AI decisions. Explainability must be defined and implemented not only from a legal perspective but also from a technical standpoint. Without concrete technical guidelines, the right to an explanation runs the risk to remain a vague and unenforceable concept.

One possible solution is the mandatory use of recognized explainable AI (XAI) methods. AI providers should be required to employ a certified XAI methodology to generate explanations that meet established transparency and comprehensibility standards. This would ensure that affected individuals receive meaningful and understandable explanations rather than generic or overly complex technical descriptions.

Additionally, standardized technical evaluation procedures should be introduced. Regulatory authorities could conduct regular audits of AI explanation mechanisms to verify their effectiveness and quality. These audits would help maintain consistent standards and prevent companies from providing insufficient or misleading explanations. Implementing such measures would enhance transparency, strengthen legal certainty, and ensure that the right to an explanation in AI decision-making is both meaningful and enforceable.

Table 1: Overview of the Key Provisions.

Fundamental Rights	
Article 8 CFR	Guarantees the right to the protection of personal data
Article 21 CFR	Prohibits discrimination based on a wide range of criteria
Article 47 CFR	Guarantees the right to an effective remedy and to a fair trial
Article 2(1) and 1(1) of the German Constitution	Guarantees the general right of personality
General Data Protection Regulation (GDPR)	
Article 22(1) GDPR	Prohibits decisions based solely on automated data processing
Article 22(2)(b) GDPR	Provides an exception to the prohibition of Article 22(1) GDPR
Article 22(3) GDPR	In the case of an automated decision, it requires the implementation of suitable measures to safeguard the data subject's rights and freedoms and legitimate interests
Article 15(1)(h) GDPR	Provides a special right of access for data subjects in cases where decisions based solely on automated processing have legal effects on them or significantly affect them in a similar manner
Article 13(2)(f) and 14 (2)(g) GDPR	Contain obligations for data controllers
Consumer Credit Directive (CCD)	
Article 18(1) CCD	Contains obligations to assess the creditworthiness of the consumers
Article 18(8)(a) CCD	Contains the right to request and obtain a clear and comprehensible explanation of the assessment of creditworthiness from the creditor - takes precedence over the right of access under Article 15(1)(h) of the GDPR - covers a broader range of automated decisions than the GDPR

AI Act	
Article 86(1) AI Act	Provides a right to explanation and applies to the high-risk AI systems
Article 86(3) AI Act	Determines the *residual* character of Article 86 (1) AI Act, meaning it applies only when no equivalent right is provided under other Union law provisions
Federal Data Protection Act (FDPA)	
§ 37a FDPA	Designed as an exception to the prohibition under Article 22(1) of the GDPR
§ 37a(4) FDPA	Regulates the disclosure obligations of controllers in the context of scoring

References

Buck-Heeb, Petra (2023): "Rechtsunsicherheiten beim KI-Einsatz im Bankensektor", in: Wertpapiermitteilungen 77(35), pp. 1625–1636.

Coalition Agreement 2021–2025 between the Social Democratic Party of Germany (SPD), ALLIANCE 90/THE GREENS, and the Free Democrats (FDP), available at: https://www.fdp.de/sites/default/files/2021-11/Koalitionsvertrag%202021-2025_0.pdf, last access: July 30, 2025.

Dubovitskaya, Elena and Buchholz, Annika (2023): "Die Geschäftsleitung und der Rat des Algorithmus", in: Zeitschrift für Wirtschaftsrecht 44(2), pp. 63–73.

Dubovitskaya, Elena and Bosold, Gregor (2024): "Die Schufa, der EuGH und das Recht auf Erklärung – Zugleich Besprechung von EuGH v. 7.12.2023 – C-634/21", in: Zeitschrift für Wirtschaftsrecht 45(32), pp. 1805–1816.

Edwards, Lilian and Veale, Michael (2017): "Slave to the Algorithm? Why a 'Right to an Explanation' Is Probably Not the Remedy You Are Looking For", in: Duke Law & Tech. Review 16, pp. 18–84.

Kumkar, Lea Katharina and Roth-Isigkeit, David (2020): "Erklärungspflichten bei automatisierten Datenverarbeitungen nach der DSGVO", in: Juristenzeitung 75(6), pp. 277–286.

Langenbucher, Katja and Bauer, Kevin (2025): "Erklärung von Kreditscores – Der Europäische Gerichtshof urteilt über automatisierte Bonitätsprüfungen", SAFE Finance Blog, May 28, https://safe-frankfurt.de/de/aktuelles/safe-finance-blog/details/explaining-credit-scores-the-european-court-of-justice-rules-on-automated-credit-assessments.html, last access: July 30, 2025.

Lundberg, Scott M. and Lee, Su-In (2017): "A Unified Approach to Interpreting Model Predictions", in: Advances in Neural Information Processing Systems 30, pp. 1–10.

Merkle, Marieke Luise (2024): "Transparenz nach der KI-Verordnung – von der Blackbox zum Open-Book", in: Recht Digital, pp. 414–420.

Metikoš, Ljubiša and Ausloos, Jef (2025): "The Right to an Explanation in Practice: Insights from Case Law for the GDPR and the AI Act", in: Law, Innovation and Technology 17(1), pp. 1–36.

Ribeiro, Marco T., Singh, Sameer and Guestrin, Carlos (2016): "'Why Should I Trust You?' Explaining the Predictions of Any Classifier", in: KDD '16: Proceedings of the 22nd ACM SIGKDD International Conference on Knowledge Discovery and Data Mining, pp. 1135–1144.

Schufa (2024): "SCHUFA arbeitet an neuer Score-Generation", May 21, https://www.schufa.de/newsroom/bonitaet/neuer-schufa-score-faq/, last access: July 30, 2025.

Seng, Leonie (2018): "Mein Haus, mein Auto, mein Roboter? ", in: Rath, Matthias, Krotz, Friedrich and Karmasin, Matthias (eds.), Maschinenethik. Normative Grenzen autonomer Systeme, Wiesbaden: Springer VS, pp. 57–72.

Wachter, Sandra, Mittelstadt, Brent and Floridi, Luciano (2017): "Why a Right to Explanation of Automated Decision-Making Does Not Exist in the General Data Protection Regulation", in: International Data Privacy Law 7(2), pp. 76–99.

Wachter, Sandra, Mittelstadt, Brent and Russell, Chris (2017): "Counterfactual Explanations Without Opening the Black Box. Automated Decisions and the GDPR", in: Harvard Journal of Law & Technology 31(2), pp. 841–887.

Weber, Robert, Kiefner, Alexander and Jobst, Stefan (2018): "Künstliche Intelligenz und Unternehmensführung", in: Neue Zeitschrift für Gesellschaftsrecht, pp. 1131–1136.

Wischmeyer, Thomas (2018): "Regulierung intelligenter Systeme", in: Archiv des öffentlichen Rechts (AöR) 143(1), pp. 1–66.

Legal Resources

Bäcker, Matthias (2024), in: Kühling/Buchner/Bäcker, Datenschutz-Grundverordnung BDSG, 4. Auflage, Art. 15 DSGVO.

BGH 28.1.2014 – VI ZR 156/13, NJW 2014(17), pp. 1235–1238.

Bienemann, Linda (2022), in: Sydow/Marsch, Datenschutz-Grundverordnung BDSG, 3. Auflage, Art. 15 DSGVO.

Buchner, Benedikt (2024), in: Kühling/Buchner/Bäcker, Datenschutz-Grundverordnung BDSG, 4. Auflage, Art. 22 DSGVO.

CJEU 29.1.2008 – C-275/06 (Promusicae), EU:C:2008:54.

CJEU 29.7.2019 – C-516/17 (Spiegel Online), EU:C:2019:625.

CJEU 21.6.2022 – C-817/19 (Ligue des droits humains), ECLI:EU:C:2022:491.

CJEU 4.5.2023 – C 487/21 (Austrian Data Protection Authority), ECLI:EU:C:2023:369.

CJEU 7.12.2023 – C-634/21 (SCHUFA Holding (Scoring)), ECLI:EU:C:2023:957.

CJEU 4.10.2024 – C-446/21 (Schrems), ECLI:EU:C:2024:834.

CJEU 27.2.2025 – C-203/22 (Dun & Bradstreet Austria), ECLI:EU:C:2025:117.

Dix, Alexander (2019), in: Simitis/Hoparaung/Spiecker gen. Döhmann, Datenschutzrecht, 1. Auflage, Art. 15 GDPR.

FCC 6.11.2019 – 1 BvR 16/13 (Right to be forgotten I), ECLI:DE:BVerfG:2019:rs20191106.1bvr001613.

Hartmann, Sarah (2024), in: Martini/Wendehorst, Verordnung über künstliche Intelligenz, Art. 86 KI-VO.

Kamlah, Wulf (2023), in: Plath et al., DSGVO/BDSG/TTDSG: Kommentar zu DSGVO, BDSG, TTDSG, 4. Auflage, Art. 13 DSGVO.

Opinion of Advocate General Pikamäe, delivered on March 16, 2023 – C-634/21 (SCHUFA Holding (Scoring)), ECLI:EU:C:2023:220.

Opinion of Advocate General de la Tour, delivered on September 12, 2024 – C-203/22 (Dun & Bradstreet Austria), ECLI:EU:C:2024:745.

Parliament of the Federal Republic of Germany 27.03.2024 – Printed Matter 20/10859, p. 1–40.

Schulz, Sebastian (2022), in: Gola/Heckmann Datenschutz-Grundverordnung, 3. Auflage, Art. 22 DSGVO.

Spindler, Gerald/Horváth, Anna Zsófia (2019), in: Spindler/Schuster, Recht der elektronischen Medien, 4. Auflage, Art. 22 DSGVO.

Cross-Chain Governance

Florian Möslein, Michael Birkner

Abstract *The governance of blockchain-based decentralized autonomous organisations (DAOs) is shaped, in part, by the architecture of the underlying blockchain networks. This chapter examines the governance challenges posed by cross-chain blockchains – advanced technologies that enable interoperability between otherwise discrete blockchain ecosystems. Building on the normative and conceptual frameworks developed in earlier chapters on digital governance and trust in AI, we argue that effective cross-chain governance requires more than technical solutions. It must also address legal coordination, participant evaluative capacities, and the design of systems capable of fostering trust. Cross-chain technologies introduce novel complexities in managing decentralized interactions across heterogeneous networks, necessitating governance models that uphold standards of security, scalability, and decentralisation. Beyond these technical criteria, we argue that legitimate and trustworthy digital governance must also incorporate normative foundations such as agency, responsibility, and practical freedom.*

1. Introduction

As blockchain technology matures, governance models are increasingly shifting from isolated, single-chain systems to complex, cross-chain ecosystems. These ecosystems are often organized through decentralized autonomous organizations (DAOs), which enable collective decision-making through smart contracts and distributed infrastructure. While DAOs promise radical decentralization, they also depend on complex forms of coordination and trust – especially when operating across multiple blockchain networks.

1.1. Challenges regarding Cross-Chain Governance

Cross-chain governance presents new challenges: synchronizing decisions across technical boundaries, maintaining security across different architectures, and ensuring compliance with legal frameworks not designed for decentralized systems. These complexities require more than technical solutions – they require a rethinking of governance itself. This chapter addresses these challenges by situating cross-chain DAO governance within the broader conceptual framework of digital governance developed in previous chapters. We draw on two complementary perspectives: the enabling capacities approach, which emphasizes the conditions under which individuals and organizations can meaningfully evaluate and engage with digital systems; and a virtue-based account of trust in technology, which shifts the focus from mere system reliability to the normative qualities embedded in the design and operation of technical infrastructures.

1.2. Conditions for Effective Governance

Applying these perspectives to the governance of cross-chain DAOs, we argue that effective governance must do more than ensure interoperability and legal compliance. It must also support participants' ability to understand, evaluate and shape governance processes – and foster trust by embedding norms of responsibility, transparency, and fairness in technical design. Our analysis thus frames cross-chain governance as a case study in applied digital governance. We explore how autonomy, legal accountability, and technological integration can be balanced in a way that preserves practical freedom and allows for normative evaluation. To this end, we examine both the technical mechanisms that underpin cross-chain interoperability and the legal challenges these systems face, drawing in particular on analogies to corporate group law within the German legal tradition (*Konzernrecht*). By integrating technical, legal, and ethical perspectives, this chapter contributes to a more holistic understanding of how decentralized systems can be governed across blockchain boundaries.

2. The Governance Model in DAOs

The governance model in DAOs is a system that allows participants to collectively make decisions about the organization's operations, direction, and policies without centralized leadership (Wright 2021). The decision-making process is not an isolated on-chain-phenomenon; rather, it must be integrated into the existing off-chain legal framework. In this regard, it falls upon the law to consider the actual intentions of the parties in question as far as is feasible within context of legal categorization.

2.1. Legal Qualification of DAOs

As opposed to traditional organizations and companies, however, the legal status of DAOs remains a complex and evolving issue, largely dependent on the jurisdiction in which they operate (for a broad comparative overview, see the contributions in Perestrelo de Oliveira and Rolo 2021). DAOs are digital organizations governed primarily by smart contracts and decentralized decision-making processes, making their legal recognition challenging under traditional legal frameworks. In most jurisdictions, DAOs acquire legal personality by default as partnerships, even without formal registration. This classification grants DAOs the ability to enter into contracts and own assets but also exposes their members to potential legal liabilities (Mienert 2022: 116 et seqs.). The absence of limited liability is a significant drawback, as participants in non-wrapped DAOs – those that do not register under a specific legal framework – may face unlimited personal liability for the organization's obligations (Möslein and Ostrovski 2024: 109 et seq.).

2.1.1. "Wrapped" DAOs and Personal Liability

To mitigate this risk, many DAOs choose to be "wrapped" by incorporating as a legal entity, such as a Limited Liability Company (LLC) or a foundation, in jurisdictions that provide a suitable regulatory framework. Some US federal states, particularly Wyoming and Vermont, have introduced specialized DAO LLC structures that provide limited liability while maintaining the decentralized governance principles of DAOs (Guntermann 2024: 480). Similarly, jurisdictions like Switzerland and the Cayman Islands offer foundation structures that can serve as legal wrappers for DAOs.

Recent legal cases highlight the risks DAO participants face due to the lack of clear legal structures. For instance, U.S. courts ruled that token holders participating in governance decisions could be held personally liable (Samuels v. Lido DAO, 3:23-cv-06492, (N.D. Cal.), see also Commodity Futures Trading Commission v. Ooki DAO, 3:22-cv-05416, (N.D. Cal.)), reinforcing the argument that DAOs need clearer legal frameworks to ensure regulatory compliance and limit liability exposure.

2.1.2. DAOs in Decentralized Finance

Beyond their role as a nexus for contracts, DAOs are increasingly relevant as a nexus for regulation (Möslein and Ostrovski 2024: 97 et seq.), particularly in the context of Decentralized Finance (DeFi). In financial markets, obtaining a license to operate legally often requires a recognized legal entity with accountable representatives. Since many jurisdictions do not allow partnerships to obtain financial service licenses, DAOs face significant regulatory hurdles. Furthermore, financial supervision laws typically assume centralized management structures, creating additional challenges for DAOs that operate through decentralized governance.

2.2. Decision-making and Governance in a Decentralized Environment

As a consequence, decision-making and governance in DAOs are fundamentally different from traditional corporate structures. DAOs operate through blockchain-based smart contracts that automate governance processes and ensure decentralization (Möslein 2020: 898 et seq.). Unlike conventional organizations, which rely on hierarchical management, DAOs distribute voting- power among their members, typically through governance tokens.

2.2.1. Token-based Voting Mechanism

DAO governance is based on a voting mechanism where members, holding governance tokens, vote on proposals related to the organization's operations. These tokens function similarly to shares in traditional companies, providing voting rights and sometimes financial benefits (Bauer 2025: 42 et seqs.). Governance token holders can vote on key issues such as protocol upgrades, resource allocation, and strategic directions.

Voting power in DAOs is often proportional to the number of governance tokens held by a participant. While this ensures that those with a larger stake have more influence, it can also lead to power centralization among wealthy participants, a challenge sometimes referred to as "whale dominance". Some DAOs attempt to mitigate this issue by introducing quadratic voting, delegation systems, or token-weighted participation thresholds (Axelsen, Jensen and Ross 2025: 77).

2.2.2. Dual-layer or Direct Token-based Voting Mechanism

The governance structure of DAOs varies widely. Some DAOs, like MakerDAO, implement a dual-layer governance system that includes both off-chain discussions (via forums and social media) and on-chain voting mechanisms. In this model, informal proposals are debated before being formally put to a vote through smart contracts. Other DAOs, such as Compound and Uniswap, use direct token-based voting where governance proposals are immediately subjected to on-chain voting (Bauer 2025: 3 et seq.).

2.3. Governance of Cross-Chain DAOs

DAOs and cross-chain governance interact by enabling decentralized decision-making across multiple blockchain networks, allowing DAOs to operate beyond a single blockchain ecosystem. As many DAOs manage assets and protocols on different blockchains, cross-chain governance mechanisms ensure that decisions made by token holders can be executed across various networks. Technologies such as cross-chain bridges, interoperability protocols (e.g., Polkadot, Cosmos), and multi-signature wallets facilitate secure communication and execution of governance actions across chains (see for instance, McCarthy 2022: 18). Smart contracts on different blockchains can synchronize DAO votes and proposals, ensuring seamless coordination. However, cross-chain governance also introduces challenges such as security risks, synchronization delays, and fragmentation of voting power, necessitating innovative solutions like oracle-based verification and decentralized interoperability frameworks.

3. Technological Background Behind Cross-Chain Interoperability

As the number of blockchains in use continues to grow, the necessity for interoperability between these disparate systems also increases. A blockchain system by its very nature operates in isolation, resulting in the so-called "blockchain-islands" phenomenon (Zhu, Zhang and Tao 2024: 1; Ou et al. 2022: 1; Li and Zhao 2024: 12007). Differences in architecture, data structure, security mechanism, or contract execution hinder the circulation of information among systems (Fu et al. 2024: 71). The creation of a unified, user-friendly ecosystem in which various blockchain platforms are interconnected, enabling the seamless performance of transactions and access to services across multiple networks without friction, represents a significant challenge for system developers. Cross-chain technology has been developed with the objective of facilitating efficient communication mechanisms between blockchains. Besides the exchange of digital assets or data between blockchains, cross-chain interoperability improves the scalability of blockchain systems.

From a technical perspective, four core strategies for cross-chain interoperability have been established: (1.) notary schemes, (2.) sidechain/relay, (3.) hash-locking and (4.) distributed private key control (Ou et al. 2022: 4). All technologies have in common that they solve the island phenomenon to a certain degree taking into account the respective limitation of each approach.

3.1. Notary Schemes

Notary schemes introduce one or more trusted third-parties to verify information between cross-chain transactions (Li, Wu and Cui 2023: 150; Ou et al. 2022: 4; similar to the role of notaries in cross-chain transactions see the role of connectors in interledger payments in: Thomas and Schwartz 2015: 2). The notary acts as an intermediary on both chains to monitor events or information and if necessary, responds on the respective blockchain (Li, Wu and Cui 2023: 150).

3.1.1. Overview: Simplified Transaction Procedure in Notary Schemes

Depending on the implemented scheme, the mechanism requires one single notary or a group of notaries (Ou et al. 2022: 4). When a transaction is conducted, a node, a group of nodes, or another entity acting as notary collects the data, verifies and confirms the transaction in the network (Ou et al. 2022: 4; on the basis of the Interledger project see Xiong et al. 2022: 1060). A node is one of the network participants that collectively run the blockchain's software. It enables the blockchain to validate transactions and keep the network secure ensuring the decentralization of the network. By using the notary scheme, the network significantly relies on the honesty and trustworthiness of each notary. The number of notaries required, the selection mechanism and consensus mechanism to be employed prior to the execution of a transaction can be individually determined by the participants in the network. Three main notary schemes have been established which differ in terms of the used signature method.

3.1.2. Single-Signature Notary Schemes

The single-signature notary scheme (also: centralized notary scheme) is the simplest of methods and delegates all notarial tasks to one single node or entity (Li, Wu and Cui 2023: 150; Ou et al. 2022: 4). In the absence of complex proof mechanisms relying on one notary may accelerate transaction times depending on the responsiveness of the single notary (Ou et al 2022: 4; Xiong et al. 2022: 1060). By entrusting solely one notary, the network significantly changes the way trust is established on blockchains. A system based purely on the distinctive characteristics of blockchain technology ("technological trust") is then supplemented by trust in the individual notary ("personal trust") in the case of cross-chain transactions. The simplicity of the single signature scheme is accompanied by the hazards of a single point of failure and a conflict with the concept of decentralization which in fact is inherent to the blockchain (Xiong et al. 2022: 1060; Mao et al. 2023: 45532).

3.1.3. Multi-Signature Notary Schemes

Multi-signature notary schemes avoid the risks of the single-signature scheme by delegating the notarial tasks to a randomly selected number of notaries out of a notary group. By giving each notary of the group a key to collusively confirm any cross-chain transaction, a higher degree of security and decentralization is achieved (Wu et al. 2023: 3; Ou et al. 2022: 4). In case of a malicious attack on some nodes or entities, the operation of the remaining system is not necessarily affected. The confirmation of the transaction depends on a certain percentage of notaries jointly signing and reaching consensus (Ou et al. 2022: 4).

Figure 1: Multi-signature Notary Mechanism (Ou et al. 2022: 4).

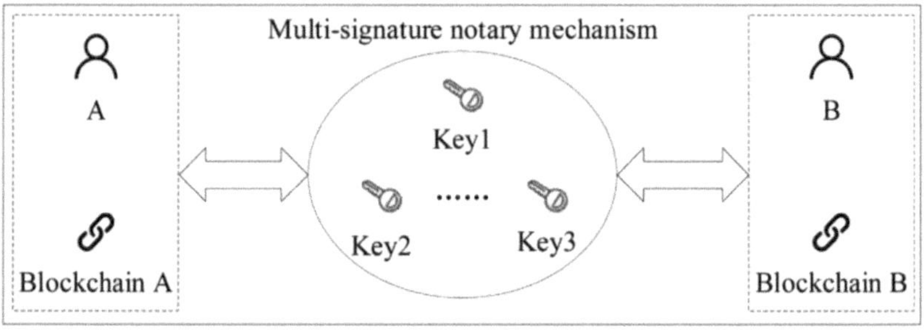

3.1.4. Distributed Signature Notary Schemes

The distributed signature notary scheme intensifies the level of security and reliability compared to the multi-signature scheme. Also avoiding centralization through the utilization of notary groups it adds a layer of security by adopting a multi-party computation (MPC) mechanism (Li, Wu and Cui 2023: 150; Ou et al. 2022: 4). The MPC is a cryptographic technique that allows multiple parties to jointly calculate an output from their private inputs while ensuring the privacy of the participants and used data (Evans, Kolesnikov and Rosulek 2020: 7 et seq.). In context of the distributed signature notary scheme, MPC is used to securitize the notary key needed to verify data and transactions. The signature key is split into multiple fragments and distributed between notaries (Ou et al. 2022: 4). The signature is only considered complete if a specific threshold of key fragments is achieved.

Figure 2: Distributed Signature Notary Mechanism (Ou et al. 2022: 5).

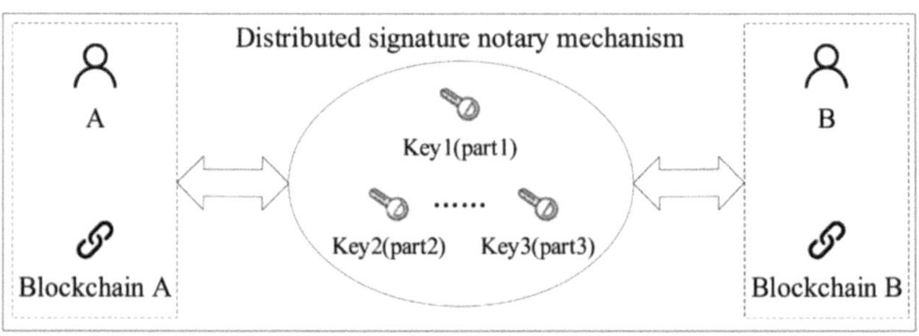

3.2. Sidechain/Relay

Sidechain and relay are two distinct cross-chain mechanisms which although often used together differ in their characterizing purpose.

3.2.1. Communication Between Blockchains via Sidechain

The sidechain is an independent blockchain linked to another blockchain. The two blockchains may be either two existing standalone blockchains, in which case they are to be treated as equals, or one blockchain may derive directly from the other (Gaži, Kiayias and Zindros 2019: 139) The term sidechain is ambiguous. A subordination relationship between both chains is not mandatory. Sidechains allow multiple blockchains to communicate with each other and react to events in the other (ibid.: 139). By use of a two-way pegging mechanism (Ou et al. 2022: 5) they allow for transfer of assets between side- and mainchain (Ou et al. 2022: 5). In principle, assets can be temporarily locked on the mainchain with subsequent release on the sidechain and vice-versa (Ou et al. 2022: 5). The assets on either chain are released once a certain number of nodes verify the locking of asset on the other blockchain. Through sidechain developers may increase the functionalities of existing and unalterable blockchain protocols by outsourcing or expanding certain functions to another blockchain (Xiong et al. 2022: 1060).

3.2.2. Relay as a Translation Tool between Blockchains

A relay facilitates cross-chain communication by providing a unified language to isolated blockchains. The purpose is limited to observation, collection, and verification of data on blockchains (Fu et al. 2024: 71). Relay builds an additional operational layer in a blockchain effectively by using smart contracts who are able to take on and verify information from other blockchains (With a relay example for Bitcoin or Ethereum see Frauenthaler et al. 2020: 2). In contrast to sidechain relays are implemented directly on the blockchain (Mao et al. 2023: 45533).

3.3. Hash-locking

3.3.1. Functionality of Hash-locks

The principle of hash-locking is based on the cryptograph technique of hash-locks in combination with a time-lock mechanism. The use of hash-locking enables the locking of assets on a blockchain via smart contracts. The lock is connected to the hash which can only be accessed by providing the correct key (Li and Zhao 2024: 12011). Time-lock mechanisms ensure that in the event of one party failing to claim their asset via key within the specified time frame, the assets of both parties in question are returned to their sender (Ou et al. 2022: 4; Li and Zhao 2024: 12011; Li, Wu and Cui 2023: 151). In this context cross chain transactions are called "atomic-swaps" meaning that the transaction is either successfully executed between both parties or no exchange of assets takes place at all.

3.3.2. The Hash – a Digital Fingerprint of Data

The hash (H) is a unique sequence of numbers and letters that represent the result of applying a cryptographic hash function to an arbitrary amount of data (Kaulartz 2016: 475). The uniqueness of the hash follows from its determinism. This means that the input of the same amount of data will always generate the same hash (Schlatt et al. 2016: 8). The amount of data (M) used in the calculation of the aforementioned hash-value represent the key to unlock the hash-lock. By sharing M, the sender enables the receiver to access the data or asset locked by H on the other blockchain.

3.3.3. Example: Cross-Chain Transaction Using Hash-locks

Example of cross-chain transaction: The sender with an asset on Blockchain A wants to exchange assets with receiver who holds an asset on Blockchain B. Typically, an exchange of assets between Blockchain A and Blockchain B via hash-locking requires 5 steps:

1) Sender generates, based on amount of data (M), a hash (H) and sends H to receiver.
2) Sender locks asset on Blockchain A through the use of H and a time-lock.
3) Receiver locks asset on Blockchain A through the use of H and a time-lock.
4) Sender can use M to unlock asset of receiver on Blockchain B.
5) By unlocking asset of receiver with M, receiver obtains M through the system and can in turn unlock asset of sender on Blockchain A.

Figure 3: Flowchart of Cross-chain Transaction Using Hash-locking (Li and Zhao 2024: 12012).

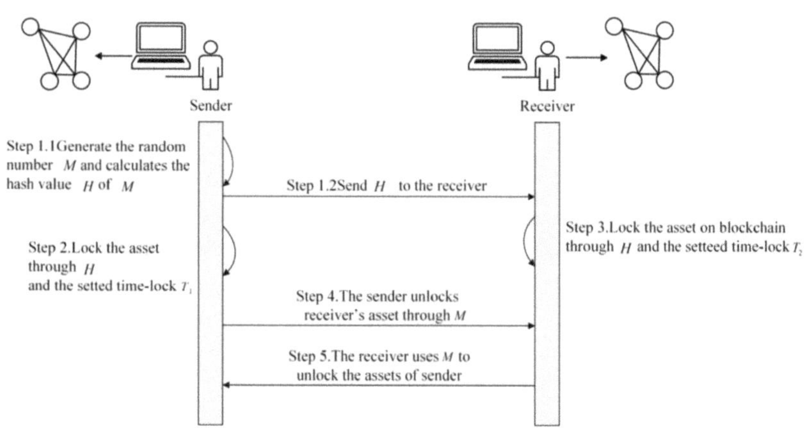

In the event that either the sender or the receiver is unable to unlock the asset on either Blockchain within the specified time frame, the cross-chain transaction collapses in on itself. In order to gain actual access to the asset, it is not sufficient to open the lock with M. Access is granted simultaneously when both parties have successfully completed the unlocking process for the asset.

3.3.4. Advantages and Disadvantages

In comparison to the notary scheme no intermediaries are required for security reasons. Due to the "all or nothing"-approach of the cross-chain transaction no trust is required between parties. Hash-locking has the advantage of low transaction costs but is limited to a specific use-case (Li and Zhao 2024: 12012; Li, Wu and Cui 2023: 151). The transaction is limited to an exchange of assets, a transfer of assets between blockchains is not possible. The total number of assets on the respective blockchain remains identical throughout the entire transaction process.

3.4 Distributed Private Key Control

Distributed private key control refers to a validation mechanism for transactions between different blockchains. Validation is required for the purpose of asset locking on the source blockchain, the creation of an equivalent asset on the target blockchain or reverse transfers of both. The validation method operates similar to the distributed signature notary scheme (see Fig. 2) by splitting the private key into fragments and distributing them across the network. By using distributed secret key generators involving the entire network of nodes a higher degree of decentralization is achieved compared to a mechanism selecting a predefined number of nodes/notaries (Ou et al. 2022: 6). During the validation process no trust between the transaction parties or any intermediary facilitating the transaction is required. Each node only saves parts of the private key and does not transmit or assemble any fragments of the whole private key. Thereby any central authority is precluded from acquiring full control over the assets (Yu and Zhang 2023: 638).

A transaction has to be confirmed on both blockchains by a number of nodes. The required validation quorum of key fragments can be defined by the protocol developer in advance (Mao et al. 2023: 45533). No single entity has full control over the asset (Yu and Zhang 2023: 638; Ou et al. 2022: 6). Ownership of the asset and the right of use fall apart (Yu and Zhang 2023: 638; Ou et al. 2022: 6). A transfer of assets or any other process may always require the authority of more than the original owner.

4. Interconnection of Blockchains

Through the interconnection of different blockchains, the user has access to several systems with different practical applications. The use of different blockchains raises the question of whether users can now expect a duplication of the governance structure, or whether interoperability as a connecting element between the chains will lead to the establishment of a single governance structure adapted to the new area of application.

4.1. Governance in Multi-Layer Systems

Cross-chain governance in DAOs does not necessarily lead to a direct duplication of their governance structure, but it often results in a more complex, multi-layered system. Since DAOs operate across multiple blockchain networks, they must establish mechanisms to coordinate decision-making and execute governance actions on different chains. This can lead to parallel governance frameworks, where each chain has its own governance process, yet remains interconnected with the broader DAO ecosystem. Some DAOs address this by using interoperability protocols, multi-chain governance tokens, or cross-chain bridges to ensure that decisions made on one blockchain are recognized and enforced on others. While these solutions promote consistency, they can also introduce inefficiencies, such as latency in governance execution and potential discrepancies in decision enforcement across chains.

4.1.1. Unification of Governance Structures

Despite these challenges, many DAOs strive to maintain a unified governance structure by implementing interoperability solutions that allow for seamless communication between different blockchain environments. Instead of fully duplicating governance processes, DAOs may adopt a federated or hierarchical model, where a primary governance layer oversees cross-chain operations while individual chains handle localized decisions. This approach minimizes redundancy while maintaining decentralized decision-making. However, achieving full coordination remains a technical and organizational challenge, requiring innovative smart contract designs, oracle-based verification systems, and standardized cross-chain protocols to ensure governance integrity and efficiency across multiple blockchain networks.

4.1.2. Legal Challenges

Cross-chain governance in DAOs also raises significant legal challenges, particularly regarding jurisdiction, regulatory compliance, and liability. Since DAOs operate across multiple blockchain networks, determining which legal framework applies to their governance decisions becomes complex. Different jurisdictions may have conflicting regulations on smart contracts, securities laws, and financial oversight, creating uncertainty for DAO participants. Additionally, enforcing governance decisions across chains can be legally ambiguous, as smart contract execution may not always align with traditional legal enforcement mechanisms. Liability is another key concern – if governance decisions lead to financial losses or regulatory breaches, it remains unclear whether responsibility falls on individual token holders, developers, or the DAO itself. As DAOs expand their cross-chain presence, legal clarity will be essential to ensure regulatory compliance while preserving decentralization and interoperability.

4.2. The Company Statute ("*Gesellschaftsstatut*") in the Cross Chain Context

In case the cross-chain model not only involves bridging the gap between different blockchains but also involves various jurisdictions it raises the question which national law system applies to the relevant legal issue (e.g. incorporation rules, liability regime, registration obligations). The applicable law in each case crucially determines the governance structure. According to German law, specifically Art. 3 EGBGB resolves this conflict of jurisdictions by applying particular rules for international private law conflicts. Sources of these rules are international treaties, the law of the European Union and German conflict of law principles which themselves are based on customary law (Mienert 2022: 83 et seq.). The international treaties and EU law prevail over national legal principles. The overriding objective of all the rules is to choose the jurisdiction with the closest connection to the subject matter.

4.2.1. Real Seat Theory ("*Sitztheorie*") or Incorporation Theory ("*Gründungstheorie*")

Due to the digital structure of cross-chain models, with the difficulty of accurate geographical allocation of users, hardware, or area of business it seems appropriate to look for a general point of contact to determine the statue of the company (for DAOs see Mienert 2022: 86). From a European perspective the real seat theory ("*Sitztheorie*") and incorporation theory ("*Gründungstheorie*") is often discussed in this context.

The idea of real seat theory connects the actual corporate seat (e.g. head offices) to the applicable company law. According to the theory, a conflict of jurisdictions is usually to be settled within the framework where the main administration is located. That sounds reasonable as it can be expected that the majority of stakeholders will also be located in this jurisdiction.

The incorporation theory determines the applicable law by reference to the jurisdiction where the company was incorporated and registered. This approach enhances legal certainty because the determination of where a company was established can usually objectively be established.

4.2.2. Application to Cross-Chain Models

Considering both theories under the organization structure of cross-chain models, significant difficulties arise in the context of applicability. Due to the decentralized nature of the chosen structure, there is no universally applicable headquarter with regard to the real seat theory.

The incorporation theory does also not help in this matter. With regard to this theory the applicable law is determined by the place of incorporation, the place of the registered office under the articles of association, the place of registration, the place chosen by the company founders (*freie Rechtswahl*) or the place under whose law the company is organized (Mienert 2022: 91). Cross-chain structures are typically built within a complete digital infrastructure without formal documentation (with the White Paper as an exception to this general rule) relying solely on its source code without references to any jurisdiction or applicable law. To determine a link between these factual levels and the legal requirements seems not useful.

4.2.3. Conflict of Chain Rules

As noted above, it is difficult to link traditional conflict of law theories to the determination of the company statue because of the special factual structures of cross-chain models. One solution could be for future cross-chain models to provide more guidance on the applicable law in the White Paper. Another option is to extend the range of connecting factors regarding jurisdiction. In addition to the country of incorporation or head office, it may be possible to resolve jurisdictional issues by focusing on the location of the main assets or servers of the cross-chain network. Even simpler, the location that is disclosed in the imprint or legal notice on the homepage of the respective project could also be considered as a reasonable linking point (Mienert 2022: 95 et seqs.).

4.3. Balancing Hierarchy and Autonomy: Parallels to Group Law ("*Konzernrecht*")

The legal challenges arising from the cross-chain governance of DAOs are primarily related to the tension between hierarchy and autonomy: On the one hand, the aim is to achieve hierarchical control across individual networks, but without giving up the advantages of decentralized autonomy. The connection and interoperability established between the different blockchains raises the question of how to manage the actual and legal relationships between them. The coming together of two previously autonomous structures leads to an area of friction that is particularly evident in the management of the structures. It is at the parties' discretion to establish a union of equals or a relationship of superiority and subordination. In the latter case effective measures have to be taken to ensure sufficient protection of minority rights. In this respect, cross-chain governance of DAOs shares notable parallels with the governance of corporate groups (*Konzernrecht*) in traditional company law, particularly under German law (for an overview: see Scheuch 2016).

4.3.1. Corporate Groups – Legally Independent but Economically Connected

In a corporate group (*Konzern*), multiple legally independent entities operate under a unified economic structure, often controlled by a parent company. Similarly, DAOs with cross-chain governance manage decentralized operations across multiple blockchain networks, where different chains function as interconnected yet technically distinct environments. In both cases, governance mechanisms must balance central oversight with local autonomy, ensuring coherence while respecting the distinct legal and operational contexts of each entity or blockchain. Just as corporate group law (*Konzernrecht*) addresses decision-making within a corporate group to prevent conflicts of interest and protect minority shareholders, cross-chain DAOs must navigate decentralized governance challenges, including voting synchronization, execution consistency, and accountability.

4.3.2. German *Konzernrecht* – Key Principles

From a legal perspective, German *Konzernrecht* imposes specific duties on the parent company in a corporate group to protect subsidiary companies and minority shareholders, particularly under the German Stock Corporation Act (AktG). A key principle is the so-called *Gleichordnungskonzern* (coordinated group), where legally independent entities operate under a common governance framework, akin to how DAOs coordinate decision-making across chains. If a DAO's governance structure resembles a so-called *Unterordnungskonzern* (hierarchical group), where a dominant chain or governance body exerts significant control over sub-chains, fiduciary duties and liability risks arise. In traditional corporate law, a parent company can be held liable if it exerts excessive influence leading to financial harm for subsidiaries (§ 311 AktG). Applying this to DAOs, if a primary governance layer enforces decisions across chains that negatively impact participants, questions of legal liability, fiduciary obligations, and regulatory compliance emerge. While DAOs generally lack formal legal personhood under German law, their growing complexity and resemblance to corporate structures may prompt future regulatory frameworks addressing governance accountability in cross-chain environments.

4.4. Liability Considerations in Multi Layer Systems

4.4.1. Personal Liability in Single Layer Systems

In the majority of jurisdictions, DAOs qualify as legal partnerships. The liability consequences for each token holder are significantly detrimental. Under the German legal framework each token holder is liable without limitation, jointly and severally for any kind of obligation of the DAO (Florstedt 2023: 843; Guntermann 2024: 480).

This applies for any obligation established before and during the membership as well as for a certain period of time after leaving the DAO. The individual token holder's liability therefore derives from the company's liability to third parties. The legal possibility of taking recourse against the company or another token holder is limited by practical concerns about anonymity and the lack of centralized management or responsibility structures (Guntermann 2024: 480). In order to avoid unfair outcomes and in view of the factual structure of the DAO, some authors suggest an institutional limitation (*institutionelle Haftungsbeschränkung*) of liability (Korch 2023: 2024). This case law established for specific subgroups of partnerships has not yet found its way into DAO case law.

The outlined liability system refers to a single-layer system. This means that a single DAO entity operates under cooperation with its token holders with third parties. This is different from the case of cross-chain models, which involve a multi-layered system based on a variety of blockchains that are linked together by the technological connection tool of their choice. In this case, different autonomous DAOs may choose to cooperate by linking their networks together to form a coherent multi-layer network. This creates a de facto merger between at least two independently functioning eco-systems to one cohesive system.

4.4.2. No Change in Liability due to the De-facto Merger of Systems

This de facto merger raises questions about the application of liability principles in multi-layer networks. In principle all token holders are liable for any obligations owed by their own network. This principle is followed by the question whether after the de facto merger in the multi-layer-system the token holders of network A are now responsible for network B's obligations and vice versa (transfer of liability). A transfer of liability may be possible by contractual assumption of obligations or by legal merger between both networks. Assuming such transfer of liability, the already existing liability risks for token holders will only be exacerbated. A contractual assumption of obligations appears implausible. Token holders typically seek for ways to limit, not expand, their liability exposure.

The aim of a legal merger is that two formerly independent legal entities (e.g. two networks) will not only be merged de facto but also de jure into one single legal entity. The assets of the legal entity being acquired and its liabilities will devolve to the acquiring legal entity. Under the German Transformation Act (UmwG) partnerships can only be merged if they are registered with the company register (*Gesellschaftsregister*). DAOs typically operating without formal registrations on a multinational basis will not be qualified for a merger under the UmwG. Another option is for one DAO to dissolve and all token holders to join the other DAO. Alternatively, both DAOs could dissolve and form a new joint DAO. Both alternatives do not constitute a merger within the meaning of the UmwG but practically lead to a combination of all token holders within one entity. However, any dissolution of partnerships requires a subsequent entity liquidation process or similar. All obligations of the DAO must be met by either the DAO or its token holders. Only after this liquidation process has been completed, the dissolvement of the entity can be realized. It is therefore not possible for any liabilities to be transferred to another entity.

4.4.3. Sole Responsibility and its Limits

As there is no possibility of transfer, the stated principle remains that each network and its token holders are solely responsible for their own liabilities. This is certainly the case when two existing DAOs decide to implement a cross-chain solution at a later stage. It may be different if a cross-chain solution is implemented from the beginning by will of all token holders. In this case, similar to single-layer solutions, a legally binding objective intent (*Rechtsbindungswille*) to form a single partnership using multiple layers is required (Fleischer 2024: 1508).

5. Enabling-Capacities Approach in Cross-Chain Governance

The Enabling-Capacities approach focuses on a shift in perspective from a purely objective focus on system characteristics to a combination with a subjective, individually driven approach. This addition recognizes that individuals cannot rely on system characteristics alone but need to acquire capacities that enable them to evaluate the system they are using on the basis of a subjective evaluation matrix. Exemplary, this matrix can be used with regard to the evaluation of the used technology within the network. The prerequisite for the Enabling-Capacity approach is that the token holders are provided with sufficient basic information.

In a rapidly growing blockchain ecosystem, an evaluation process from the token holder's perspective appears to be significantly cumbersome and depending on the investment volume not economical. This applies all the more when, in context of cross-chain solutions, a de facto merger can lead to a duplication of structures that need to be examined. The large number of presented technological cross-chain solutions are an exemplary for the complexity of the technology. We therefore propose that the flow of information must be centralized from the single network itself to each token holder by White Paper or prior to each individual decision by an individual letter of information. The latter may be incorporated within a smart contract by means of an if-then formula. If a network collectively decides to opt for a cross-chain solution, then all token holders have to be thoroughly informed about the possible technological solutions as well as their advantages and disadvantages. From a legal perspective, this results in a mandatory preventive fulfilment of the information rights (§ 717 BGB) of token holders. Preventative fulfilment is central for the Enabling-Capacities approach due to the lack of a central authority in the DAO meaning that legal enforcement of information rights by its token holders involve considerable practical difficulties (Guntermann 2024: 481).

6. Conclusion

Cross-chain governance represents a critical evolution in the decentralized ecosystem, enabling DAOs to operate beyond the limitations of single-blockchain frameworks. However, this expansion introduces significant challenges related to governance coordination, security risks, and legal uncertainties. Technological solutions such as cross-chain bridges, interoperability protocols, and oracle-based verification systems offer pathways to more effective governance, yet their implementation remains complex. From a legal perspective, the parallels between cross-chain DAOs and corporate group law highlight the need for regulatory clarity, particularly in ensuring accountability and protecting stakeholder interests.

6.1. Harmonization and Legal Certainty

As DAOs continue to scale across multiple blockchains, future regulatory and technological developments will be crucial in shaping a governance model that balances decentralization, efficiency, and legal compliance. Legislators across the world face a constantly changing digital ecosystem and may find it challenging to adapt their regulatory approach to an evolving system which does not wait on approval, frameworks, or jurisdictions. The accelerating interconnection of blockchains is driving the internationalization of ecosystems and increase the need for overarching, harmonizing regulations. Exemplary, in the context of public limited liability companies (*Aktiengesellschaften*), the EU legislator has ensured minimum equivalent protection for shareholders and creditors by the use of a harmonizing directive (Directive EU 2017/1132).

In addition to preventive general legislation, the judiciary can facilitate legal certainty with regard to these new governance structures by applying already existing laws through case law. However, the responsibility falls upon the private sector to initiate legal proceedings in such cases.

6.2. Sandboxes as Education Tools

Regulators, supervising authorities and private cross-chain innovators may engage in a regulatory dialogue in a safe, confidential and controlled environment. Sandboxes can help to identify barriers from a legal, regulatory and private perspective involving all relevant stakeholders. Cross-chain governance models can be tested for a limited period of time before being integrated into existing market structures.

Overall, this can help remove barriers and strike a balance between innovation and regulatory oversight, thereby improving legal certainty.

The idea of controlled testing environments at EU-level is not new and has already been established with regard to the European Blockchain Regulatory Sandbox or within the EU Artificial Intelligence Act. It seems reasonable to build on existing structures with a particular focus on cross-chain models. The integration into the European Blockchain Regulatory Sandbox with the expansion of supported projects is advisable.

References

Axelsen, Henrik, Jensen, Johannes R. and Ross, Omri (2024): "When is a DAO Decentralized?", in: Perestrelo de Oliveira and Rolo, Decentralised Autonomous Organisation (DAO) Regulation. Principles and Perspectives for the Future, pp. 59–91.

Bauer, Philipp (2025): Governance-Token. Ein dezentralisiertes und tokenisiertes Mitgliedschaftspapier?, Mohr Siebeck.

Directive (EU) 2017/1132 of the European Parliament and the Council Directive (EU) of 14 June 2017 relating to certain aspects of company law.

Evans, David, Kolesnikov, Vladimir and Rosulek, Mike (2020): A Pragmatic Introduction to Secure Multi-Party Computation, Boston, MA: Now Publishers.

Fleischer, Holger (2024): "Plötzlich Personengesellschafter. Eine kleine Phänomenologie unbeabsichtigter Personengesellschaften", in: ZIP: Zeitschrift für Wirtschaftsrecht 45(27), pp. 1501–1512.

Florstedt, Tim (2023): "Tokengesellschaftsrecht – Zur Organisations- und Vermögensverfassung digitaler Gesellschaften (Decentralized Autonomous Organizations)", in: Zeitschrift für Unternehmens- und Gesellschaftsrecht 52(6), pp. 816–848.

Frauenthaler, Philipp et al. (2020): "Leveraging Blockchain Relays for Cross-Chain Token Transfers", in: White Paper, TU Wien, pp. 1–5.

Fu, Wanshu et al. (2024): "A Blockchain Cross-Chain Solution Based on Relays", in: International Journal of Knowledge and Innovation Studies 2(2), pp.70–80.

Gaži, Peter, Kiayias, Aggelos and Zindros, Dionysis (2019): "Proof-of-Stake Sidechains", in: 2019 IEEE Symposium on Security and Privacy 1, pp. 139–156.

Guntermann, Lisa (2024): "Wyoming DUNA – ein neuer 'legal wrapper' für DAOs", in: RDi 2024, pp. 476–486.

Kaulartz, Markus (2016): "Hintergründe zur Distributed Ledger Technology und zu Blockchains", in: Computer und Recht 32(7), pp. 474–480.

Korch, Stefan (2023): "Haftung und Haftungsprivilegien in Decentralized Autonomous Organizations", in: ZIP: Zeitschrift für Wirtschaftsrecht 44(39), pp. 2017–2025.

Li, Jiao and Zhao, Wanting (2024): "Blockchain Cross-chain Protocol Based on Improved Hashed Time-locked Contract, in: Cluster Computing 27, pp. 12007–12027.

Li, Li, Wu, Jiahao and Cui, Wei (2023): "A Review of Blockchain Cross-chain Technology", in: IET Blockchain 3(3), pp. 149–158.

Mao, Hanyu et al. (2023): "A Survey on Cross-chain Technology. Challenges, Development, and Prospect, in: Ieee Access 11, pp. 45527–45546.

McCarthy, Sam (2022): Stewards and Gatekeepers. Human and Technological Agency in the Governance of DeFi Protocols, SSRN, August 18, 2022, https://ssrn.com/abstract=4326903.

Mienert, Biyan (2022): Dezentrale autonome Organisationen (DAOs) und Gesellschaftsrecht. Zum Spannungsverhältnis Blockchain-basierter und juristischer Regeln, Mohr Siebeck 2022

Möslein, Florian (2020): "A Nexus of Smart Contracts? Gesellschaftsrechtspraxis und -theorie im Spiegel der Blockchain", in: Bachmann, Gregor and Grundmann, Stefan et al. (eds.), Festschrift für Christine Windbichler zum 70. Geburtstag, de Gruyter (2020), pp. 889–904.

Möslein, Florian and Ostrovski, Daniel (2024): "Legal Personality of Decentralized Autonomous Organisations (DAOs). Privilege or Necessity?", in: Perestrelo de Oliveira and Rolo, Decentralised Autonomous Organisation (DAO) Regulation. Principles and Perspectives for the Future, pp. 93–112.

Ou, Wei et al. (2022): An overview on cross-chain: Mechanism, platforms, challenges and advances, in: Computer Networks 218(109378), pp. 1–21.

Perestrelo de Oliveira, Madalena and Rolo, António G. (eds.) (2024): Decentralised Autonomous Organisation (DAO) Regulation. Principles and Perspectives for the Future, Mohr Siebeck.

Perestrelo de Oliveira, Madalena and Rolo, António G. (eds.) (2023): Decentralised Autonomous Organisations (DAOs) in Various Jurisdictions. From Old Rules to Innovative Approaches, https://www.cidp.pt/publication/decentralised-autonomous-organisations-daos-in-various-jurisdictions-from-old-rules-to-innovative-approaches/292, last access: April 5, 2025.

Scheuch, Alexander (2016): "Konzernrecht. An Overview of the German Regulation of Corporate Groups and Resulting Liability Issues", in: European Company Law 13, pp. 191–198.

Schlatt, Vincent et al. (2016): "Blockchain. Grundlagen, Anwendungen und Potenziale", Projektgruppe Wirtschaftsinformatik des Fraunhofer-Instituts für Angewandte Informationstechnik FIT, https://www.fim-rc.de/Paperbibliothek/Veroeffentlicht/642/wi-642.pdf, last access: March 3, 2025.

Thomas, Stefan and Schwartz, Evan (2015): "A Protocol for Interledger Payments", https://interledger.org/developers/documents/interledger.pdf, last access: March 3, 2025.

Wright, Aaron (2021): "The Rise of Decentralized Autonomous Organizations. Opportunities and Challenges", in: Stanford Journal of Blockchain Law & Policy, pp. 152–176.

Wu, Xiaohua et al. (2023): "A Distributed Cross-chain Mechanism Based on Notary Schemes and Group Signatures", in: Journal of King Saud University – Computer and Information Sciences 35(101862), pp. 1–14.

Xiong, Anping et al. (2022): "A Review of Blockchain Cross-chain Technology, in: Digital Communications and Networks 8, pp. 1059–1067.

Yu, Yue and Zhang, Shibin (2023): "A Cross-Chain Identify Authentication Scheme Based on Block Chain", in: 2022 3rd International Conference on E-commerce and Internet Technology, Atlantis Press, pp. 635–643.

Zhu, Zeshuo, Zhang, Rui and Tao, Yang (2024): "Atomic Cross-chain Swap Based on Private Key Exchange", in: Cybersecurity 7(12), pp.1–22.

Blockchain-Based Governance of Financial Markets

Examining the SEC's Approach to Building Trust in Centralized and Decentralized Crypto Exchanges

Sebastian Omlor, Hans Wilke

Abstract *Crypto assets have emerged as a driving force in the modern financial economy. Contrary to their original conception as a means of payment, they are now predominantly held as investment instruments by retail investors worldwide. This growing market is fueled by decentralized platforms that enable the exchange of crypto assets. This development raises fundamental questions about the standards these platforms must meet to be considered trustworthy and how retail investors can evaluate that trustworthiness. Regulatory frameworks play a central role in addressing these challenges. This article analyzes the approach taken by the U.S. Securities and Exchange Commission under Chair Gary Gensler toward the regulation of decentralized exchanges. It concludes by offering regulatory guidelines aimed at shaping the ongoing legislative discussions in both chambers of the U.S. Congress on how to make decentralized exchanges more trustworthy.*

Over the past decade, the market for crypto assets has experienced exponential growth. As of 2024, the global market capitalization of crypto assets fluctuated around $3 trillion, up from a few billion dollars in the early 2010s (Westbrook 2024). This development is not only driven by speculative interest but also by the emergence of a liquid and increasingly complex secondary market for crypto assets. At the center of this market stand crypto exchanges, which serve as both cornerstones and gatekeepers of the infrastructure that enables the exchange of digital assets.

However, the crypto exchange landscape has been riddled with scandals, collapses, and misappropriations. The downfall of the crypto exchange FTX serving as a particularly stark example (Huang, Osipovich and Kowsmann 2022). These developments have brought to the fore a fundamental question: When can crypto issuers and crypto investors trust the systems that facilitate the issuance and trading of crypto assets?

The legal environment, defined not merely by formal rules but also by the credibility and consistency of their enforcement, is critical in fostering such trust. A robust regulatory framework mitigates risks of expropriation, fraud, and manipulation by issuers, thereby increasing investor willingness to commit capital (La Porta et al. 1997). In the secondary market, regulatory oversight of crypto exchanges can enhance transparency, integrity, and investor protection—key elements of a trustworthy infrastructure for financial marketplaces. As such, regulation has the potential to foster trust in systems that underpin digital asset markets.

Yet, regulation is not an end in itself. Trust-enhancing regulation must serve broader public interest goals. A trustworthy digital financial system must not only be normatively desirable in terms of outcomes (e.g. investor protection and capital formation), but also understandable and controllable. The normative orientation of financial regulation is thus twofold: It must promote the formation of capital and protect investors, while also ensuring that systems can be understood and evaluated.

This paper examines how the U.S. Securities and Exchange Commission (SEC) seeks to build the outlined elements of trust in the secondary markets for crypto assets. The SEC's approach has been described as "regulation by enforcement" and is characterized by applying existing regulatory frameworks to both centralized and decentralized exchanges (Kazimirov 2024).

To this end, former SEC Chair Gary Gensler had repeatedly urged digital asset exchanges to "come in and register" (Lutz 2022). These statements sent shockwaves through the crypto industry. However, they did not have the desired effect – digital asset exchanges did not register as national securities exchanges under the Securities Exchange Act of 1934. In response, the SEC pursued aggressive enforcement actions, notably against Coinbase Global, Inc. (Coinbase), one of the largest centralized exchanges in the U.S. The agency accused Coinbase of operating an unregistered securities exchange (SEC 2023). Coinbase had gone public with SEC approval just two years prior (Coinbase 2021b).

The SEC also targeted decentralized exchanges (DEXs). It proposed an amendment to the definition of "securities exchange" under the 1934 Act, seemingly aimed at encompassing decentralized trading platforms (88 Fed. Reg. 29,448, May 5, 2023). Further, the SEC issued a Wells notice to the developers of the Uniswap protocol, signaling potential violations of securities laws (Adams 2024).

The SEC's actions suggest it operates under two key premises: (1) both centralized and decentralized crypto exchanges qualify as securities exchanges under the 1934 Act, and (2) compliance with the existing regulatory framework for traditional exchanges is feasible and desirable for crypto exchanges, because it contributes to the trustworthiness of the system. This paper critically examines the second premise.

Establishing a system as trustworthy requires, first and foremost, an understanding of how that system functions. Accordingly, the first part of this paper provides a functional overview of how assets are traded in both traditional financial markets and crypto-asset markets, using Coinbase – a centralized exchange – and Uniswap – a decentralized exchange – as case studies (A). A closer look at both forms of crypto exchanges reveals that the systems used to facilitate the exchange of assets are far less opaque than commonly assumed. They draw on concepts already employed in traditional financial markets.

Based on these observations, we outline focus points of the regulatory framework applicable to traditional exchanges and assess its compatibility with centralized and decentralized crypto exchanges. We argue that compliance with the current regulatory scheme is neither feasible nor desirable for both types of crypto exchanges (B). Gensler's call for registration rings hollow. The SEC's current approach does little to enhance the trustworthiness of crypto market infrastructure, as the application of current regulations fails to realize the overarching policy objectives of capital formation and investor protection.

In Part C, we conclude by advocating for a tailored regulatory framework for crypto-asset markets. We propose a set of regulatory guardrails, designed to promote transparency, reliability, and investor trust, while aligning with broader public interest goals. These focus areas, we argue, are essential for rendering crypto asset exchanges more trustworthy.

1. Contrasting Traditional Markets for Equity Securities and Markets for Crypto Assets

1.1. Traditional Markets for Equity Securities

In the era when equity securities were primarily traded on the physical floors of stock exchanges, the process of purchasing equity securities typically followed a series of predefined steps. A retail investor wishing to acquire a security would initiate the transaction by engaging a broker-dealer – an intermediary with specialized knowledge of securities transactions. The broker-dealer, as a member of a securities exchange, would operate within the rules and regulations governing that exchange, which prescribed which securities could be traded and the procedures for matching buy and sell orders (Harris 2002: 34).

Upon receiving the investor's order, the broker-dealer would take it to the exchange, where it would be matched with a corresponding sell order from another broker-dealer representing a different investor. The details of the transaction would be reported to the exchange's associated clearinghouse. The clearinghouse would act as the counterparty to both sides of the transaction, absorbing the risk of default by the trading brokers (Armour et al. 2016: 16). The entire clearing process typically took several days to complete, after which the transaction would be considered fully settled and both brokers would report back to their respective clients.

This simplified explanation serves to highlight the numerous intermediaries involved in the traditional equity market, even for what might appear to be a straightforward exchange of stock for fiat currency. While the technology and infrastructure of modern markets have evolved significantly and trading is hardly conducted on physical trading floors, the fundamental roles of these intermediaries – broker-dealers, exchanges, and clearing agencies – have remained constant.

1.2. Decentralized and Centralized Exchanges for Crypto Assets

A central promise of applying blockchain technology to financial markets, is its potential to eliminate the outlined intermediaries by replacing them with automated computer code (Johnson 2023; Lee 2016). This is achieved through a set of code referred to as a smart contract, which execute specific actions when predefined conditions are met. The term has often puzzled lawyers but has nothing to do with contract law. Instead, the name alludes to the concept of a set of rules (akin to contractual provisions) being executed automatically – hence, "smart." Often compared to vending machines, smart contracts automatically process transactions based on user inputs, triggering a predetermined outcome (Shahin 2025).

Applied to financial markets, smart contracts can facilitate direct asset exchanges. For example, a user looking to swap asset X for asset Y interacts with a smart contract, which verifies whether the transaction can be executed according to its predetermined rules for facilitating asset exchanges. If the conditions are met, the contract completes the exchange without taking custody of the assets, effectively removing the need for a broker-dealer, centralized exchange, or clearinghouse. The benefits of replacing intermediaries with code are apparent. Smart contracts reduce the risk of mismanagement, conflict of interest, or unauthorized trade execution, since smart contracts operate based on transparent, pre-defined rules. Additionally, eliminating intermediaries has the potential to lower transaction costs for users and speed up the timeline of the transaction (Lee 2016).

The potential of blockchain technology for transforming financial marketplaces has garnered significant attention. However, to date, blockchain technology is scarcely used in markets trading traditional equity securities. Instead, blockchain-based trading systems are more commonly applied to trading blockchain-based assets.

Blockchain-based assets are traded on both centralized and decentralized exchanges, both of which have faced regulatory scrutiny. This section of the paper examines how these platforms facilitate the exchange of crypto assets and how their methods differ from those used in traditional financial markets. It is on this basis that the compatibility of regulatory requirements for traditional financial markets can be assessed for blockchain-powered markets.

1.2.1. Centralized Crypto Exchanges: Coinbase

Coinbase facilitates the exchange of crypto assets through a centralized process. Customers create an account on the platform and deposit either fiat currency or crypto assets. To place a buy order, they must have sufficient funds to cover the transaction value plus any fees, while a sell order requires that the corresponding crypto assets be available in their account. This structure means that Coinbase takes custody of its customers' assets (Coinbase n.d.-c).

Trades on Coinbase occur through an order book system, where buy and sell orders are matched using an automated matching engine. Customers submit their orders without knowing the identity of their counterparties. The matching engine operates on a price-time priority rule, categorizing orders as either maker or taker. A maker order is a limit order that cannot be immediately matched and remains in the order book, contributing to market liquidity. A taker order, by contrast, is a marketable order that can immediately be paired with an existing maker order at the best available price. If a taker order is only partially filled, it continues matching with subsequent maker orders until it is fully executed, or no further matches are available (Coinbase n.d.-b). Coinbase incentivizes customers to provide liquidity to the market, by rewarding customers that post orders that later become maker orders (Coinbase n.d.-a).

Once a trade is executed, settlement occurs automatically through debits and credits to customer accounts. However, ownership changes are recorded solely on Coinbase's internal ledger, rather than on the blockchain. Transactions are processed off-chain, meaning no updates are made to the blockchain associated with the traded asset (SEC v. Coinbase, No. 23-cv-06438, S.D.N.Y. 2023).

This centralized model diverges significantly from the framework envisioned by blockchain technology and smart contracts. Aside from the crypto assets being traded, blockchain technology itself plays little role in Coinbase's operations. Rather than eliminating intermediaries, Coinbase consolidates their functions into a single entity, effectively acting as a broker-dealer, exchange, and clearinghouse. The centralized model effectively creates a single point of failure and does not leverage the advantages of blockchain technology and smart contracts.

1.2.2. Decentralized Exchanges: Uniswap

The Uniswap protocol facilitates the exchange of crypto assets native to the Ethereum blockchain. It allows users to swap a predefined set of tokens against each other. It employs automated market maker technology to facilitate the trading of digital assets. Instead of matching buy and sell orders submitted by users, automated market maker technology facilitates the creation of pools of crypto assets. These liquidity pools are smart contracts that hold balances of two unique tokens and enforce rules around depositing and withdrawing them.

The exchange rate within a liquidity pool is adjusted based on the ratio of tokens in the pool. The protocol follows a mathematical pricing formula that ensures that the product of token reserves remains constant (Uniswap 2021). As traders swap one token for another, the relative supply shifts, causing the price to adjust. If a token becomes scarce in the pool, its price increases relative to the other token. A highly liquid pool experiences less price impact from individual trades than a less liquid pool (ibid.).

Users that provided liquidity to the pool, do not withdraw the exact tokens they initially deposited. Instead, they receive a proportional share of the pool's total assets at the time of withdrawal. This exposes them to a phenomenon described as impermanent loss, which occurs when the relative price of the two tokens changes, after they have deposited liquidity. Because the protocol automatically rebalances token holdings, liquidity providers may end up with a greater proportion of the asset that has decreased in value. This loss is only realized if they withdraw at a point where the price divergence still exists, making it impermanent (Gemini n.d.-a).

The protocol does not take custody of the assets that swappers use (Lin 2019). To persist with our vending machine analogy, initiating a swap via Uniswap is like inserting a dollar bill into a vending machine and selecting a product. The machine will only accept the bill if it can dispense the requested product.

In theory, anyone with the technical knowledge can interact with the protocol – no intermediary is needed. However, most users do not have the required technical expertise. Instead, they access the protocol via applications that offer a more intuitive user experience. Uniswap developers offer an application that allows users to interact with the deployed protocol.

Effectively, the outlined process replaces the intermediaries in traditional financial markets with a set of fixed rules implemented through computer code. However, in most cases, an intermediary remains involved in the form of the application used to access the protocol.

2. The Regulatory Challenges of Traditional Frameworks in the Context of Crypto Assets

We examine the regulatory framework applicable to traditional securities exchanges and argue that the regime introduced by the 1934 Act is fundamentally based on intermediation, making it conceptually incompatible with decentralized exchanges and decentralized assets (I.).

We then broaden our examination to include centralized exchanges and argue that, while their structure may seem more compatible with existing regulations, the nature of the assets traded poses significant challenges. As a result, compliance with various regulatory provisions is either infeasible for both types of crypto exchanges or undesirable, as it would fail to achieve the overarching policy objectives and therefore not contribute to the trustworthiness of the system (II.). We believe our findings underscore the need for new regulation that considers the decentralized nature of crypto assets.

2.1. The Regulatory Mismatch Between Traditional Exchanges and Decentralized Exchanges

The regulatory framework governing securities exchanges is deeply rooted in the structure of equity markets, as they existed at the time of the 1934 Act. The Act was designed for a marketplace composed of broker-dealers, issuers, and exchanges. Every participant was envisioned as a centralized, tangible entity subject to a specific set of rules and regulations.

The 1934 Act envisions exchanges as mutual or cooperative entities composed of members who transact in securities. Early exchanges were established by stockbrokers seeking a dedicated marketplace – not to attract outside traders, but to facilitate transactions among themselves (Soderquist and Gabaldon 2014: 112). The New York Stock Exchange (NYSE), for instance, initially operated out of a rented room at 40 Wall Street, where brokers convened twice daily to trade a limited list of 30 stocks and bonds (NYSE 2025). Under this traditional model, brokers simultaneously served as the primary customers, owners, and managers of the exchange (Fleckner 2006). Reflecting this framework, the 1934 Act defines stock exchanges as organizations composed exclusively of broker-dealer members. Section 6(c) of the Act stipulates that only registered broker-dealers may be admitted as exchange members. This envisioned close connection between exchanges and their members is further reinforced by Section 6(b)(1), which obligates exchanges to enforce compliance with the 1934 Act and all related regulations by its members.

Similarly, the act presupposes a centralized product, issued by an identifiable issuer. Section 12(a) of the 1934 Act prohibits transactions on an exchange in a security that is not registered with the SEC. The registration process, outlined in Section 12(b), requires issuers to provide substantial information about both the issuer and the security to be traded. Additionally, Section 13(a) mandates that issuers of registered securities submit ongoing disclosures, including annual reports (Form 10-K), quarterly reports (Form 10-Q), and current reports (Form 8-K) for material events. The Act thus assumes a centralized product listed on an exchange, with a clear issuer responsible for ongoing regulatory compliance.

This structure makes sense within traditional markets, where intermediaries serve as essential gatekeepers subject to their own regulatory obligations. Crypto markets, however, operate under an entirely different paradigm. Traditional broker-dealers are largely absent, replaced instead by retail investors transacting directly on exchanges. Similarly, the role of the issuer – central to traditional securities markets – is often filled by decentralized protocols with no identifiable controlling entity.

As outlined, the 1934 Act presupposes the existence of intermediaries, making its application to decentralized exchanges and crypto assets conceptually problematic. A crypto exchange that becomes a registered exchange under the 1934 Act would have no assets to trade, given the fact that hardly any crypto assets are registered with the SEC (Committee on Capital Markets Regulation 2023). Crypto exchanges will be hard-pressed to locate Satoshi Nakamoto and require ongoing disclosures about Bitcoin. The outlined membership limitation for broker-dealers on exchanges presents a similar problem. Broker-dealers are required to register with the SEC under Section 15(a) of the 1934 Act. The SEC's rules impose specific requirements on broker-dealers operating in the cryptocurrency space, including the restriction that they can only engage in business activities related to registered crypto assets (SEC 2021). This means that broker-dealers are prohibited from handling unregistered crypto assets, which could include commodities like Bitcoin. This makes operating as a broker-dealer unattractive, especially given the uncertain regulatory classification of crypto assets as securities. When the foundational elements of intermediation are missing, the traditional regulatory framework struggles to adapt. The 1934 Act was built for a world of intermediation, whereas crypto markets are premised on disintermediation (Coinbase 2021a).

This regulatory disconnect is evident in the SEC's enforcement actions. The SEC seems to allege that Uniswap simultaneously operates as an unregistered broker-dealer, exchange, and clearing agency (Uniswap 2024). However, this approach fails to acknowledge that decentralized exchanges do not merely assume the roles of traditional intermediaries. Rather than acting as direct substitutes for broker-dealers or clearinghouses, decentralized exchanges facilitate transactions in a manner that often eliminates the need for such intermediaries altogether. In a world where trades are settled automatically by computer code, it makes little sense to require crypto exchanges to register as clearing houses and follow the regulatory regime imposed on them. The definition of a clearing house under the 1934 Act is provided in Section 3(a)(23) and appears broad enough to encompass instantaneous settlement. The SEC's approach – imposing legacy regulations designed for a fundamentally different market structure – raises serious concerns. Forcing the remaining intermediaries to comply with all regulatory obligations previously assigned to multiple intermediaries, is neither feasible nor effective.

2.2. Crypto Assets and the Infeasibility of Full Regulatory Compliance

SEC and congressional regulation have profoundly shaped the current market structure for equity securities. We seek to highlight the fundamental conceptual differences between traditional financial markets and crypto asset markets by demonstrating that many of the policy objectives underlying existing rules and regulations and implemented to foster trust in financial markets cannot be effectively achieved when applied to decentralized and centralized crypto exchanges.

For instance, the rules established under Regulation National Market System (NMS), designed to create a consolidated market for securities and ensure investors execute trades at the best possible price, are ill-suited for crypto markets (2.2.1). Similarly, SEC requirements aimed at preventing fraudulent and manipulative practices, as mandated by Section 6(b)(5) of the 1934 Act, cannot be reasonably applied to markets for crypto assets (2.2.2).

2.2.1. The Nationl Market Place for Equity Securities

Traditionally, exchanges competed for the listing of securities, but are not for trading in the same stock. A stock would only trade on the exchange it was listed on. The NYSE and National Association of Securities Dealers Automated Quotations (NASDAQ) held a near monopoly in the trading of stocks they listed (Coffee, Sale and Whitehead 2020: 654). This has changed. Regulatory intervention has led to the development of a national marketplace for equity securities listed on national securities exchanges. The emergence of this market structure can be traced back to the addition of Section 11A to the 1934 Act which gave the SEC the task of facilitating the creation of a national market system (detailed examination in Fox et al. 2018). The SEC has implemented various measures to develop the marketplace envisioned by Section 11A. Today, the national market system operates primarily under Regulation NMS and the Unlisted Trading Privileges Act.

Section 12(a) of the 1934 Act prohibits brokers and dealers from executing transactions in any security on an exchange, unless that security is registered on that specific exchange. In principle, this means a stock can only be traded on the exchange it is listed on (Beny 2002). However, Section 12(f) initially empowered the SEC to grant unlisted trading privileges for a stock listed on another registered exchange. The Unlisted Trading Privileges Act of 1994 amended Section 12(f), broadly permitting exchanges to trade stocks listed on different registered exchanges without requiring SEC approval.

The fact that a stock can trade on different exchanges, can lead to a fragmentation of the market for the individual stock. Different exchanges operating independently from each other, can offer different prices and liquidity dispersions for the same security. Congress aimed to consolidate the market by adding Section 11A to the 1934 Act. Section 11(a)(1)(C) of the 1934 Act emphasizes the public interest in ensuring fair competition among exchange, investor access to securities quotations and transaction data, and the ability of brokers to execute orders in the best available market. Essentially, this provision suggests that a broker handling a customer's buy or sell order, should have visibility into price quotations across all markets where the stock is traded and route the order to the venue offering the best price.

Under SEC Rule 602(a), exchanges must collect and process data on the best bid, best offer, and aggregate quotation size for each security traded on their platforms. Rule 603(b) then requires exchanges to collaborate under a national market system plan to consolidate this information across all exchanges. The data collected on the individual exchanges is consolidated by a Securities Information Processor (SIP) in accordance with Rule 603(b)(3). The SIP then disseminates this data to market participants. The highest and lowest offer disseminated by the SIP are known as the National Best Bid (NBB) and National Best Offer (NBO), collectively referred to as the National Best Bid and Offer (NBBO) under Rule 600(b)(60). To ensure access to a complete view of the market when trading decisions are made, Rule 603(c)(1) prohibits SIPs from providing a display of information on quotations, without also providing a consolidated display of information.

Under the Order Protection Rule (Rule 611), introduced in 2005, exchanges must ensure that trades are not executed below the NBBO at the time an order is submitted. This rule is commonly referred to as the "trade-through" rule because it aims to prevent transactions from being executed at a price worse than the NBBO – in other words, from ignoring or "trading through" the bid designated as the NBBO.

Effectively, the rules promulgated under Section 11A of the 1934 Act have transformed the US market for listed equity securities into one consolidated limit order book (Coffee, Sale and Whitehead 2020: 662).

The application of Regulation NMS to crypto exchanges presents challenges. It remains unclear whether exchanges would be required to aggregate and disseminate data on quotes and transactions for crypto assets traded on their platforms, treating them as NMS securities under Regulation NMS. Likewise, it is uncertain whether they would need to generate an NBBO and implement trade-through protections under Rule 611. From a policy standpoint, these requirements are designed for a consolidated market – something that does not exist for crypto assets. Given the low barriers to entry, the global nature of crypto trading, and its 24/7 operation, the creation of a consolidated market as envisioned by Section 11A of the 1934 Act for digital assets seems impossible. However, without a consolidated market, forcing the few registered national crypto exchanges to consolidate and disseminate their quotes would fail to provide investors with the desired comprehensive view of crypto asset prices and order execution at the best available price. Instead, it would impose excessive compliance costs on these exchanges and stifle innovation. Moreover, it would place U.S.-based crypto exchanges at a competitive disadvantage compared to offshore platforms that are not subject to similar regulatory burdens, potentially driving market activity abroad and weakening domestic oversight.

2.2.2. Prevention of Fraudulent or Manipulative Practices

Section 6(b)(5) of the 1934 Act mandates that the rules of a national securities exchange must be designed to prevent fraudulent and manipulative acts and practices, promote just and equitable principles of trade, and protect investors and the public interest. The SEC has interpreted this provision to require exchanges to refrain from listing assets if there are significant concerns about potential fraudulent or manipulative practices in the market for those assets. Exchanges are responsible for implementing effective measures to detect and prevent manipulation (an overview of the SEC's line of argument can be found in Dombalagian 2024). For assets traded across multiple markets, these measures are typically enforced through surveillance-sharing agreements (SEC 2018). In the case of Bitcoin Exchange-Traded Products (ETF), the SEC had previously denied applications due to concerns about market manipulation and the lack of a comprehensive surveillance-sharing agreement with the major markets trading the underlying Bitcoin assets. The SEC has emphasized that, to comply with Section 6(b)(5), exchanges must demonstrate the ability to prevent fraudulent and manipulative practices. It argued that for Bitcoin such compliance would be impossible, due to its global, decentralized, and largely unregulated market structure. The SEC eventually approved spot bitcoin exchange-traded products, but only after a court vacated the SEC's denial of an application by Grayscale Investments, LLC, to convert its Bitcoin Trust into a Bitcoin ETF (SEC 2024).

Again, the application of Section 6(b)(5) to marketplaces for crypto assets presents significant challenges due to the fundamentally different market structure of digital assets compared to traditional securities. Most crypto assets trade in a fragmented, global, and largely unregulated environment. A regulated U.S. exchange listing a digital asset security that is also actively traded on unregulated platforms would lack comprehensive oversight of the full market activity for that asset. Without such oversight, the exchange would be unable to effectively detect and prevent manipulation, as required by Section 6(b)(5) of the 1934 Act. Given these structural differences, even if U.S. exchanges were prohibited from listing a particular digital asset security, that asset could still be widely traded on unregulated platforms. This undermines the SEC's ability to enforce fair and orderly market conditions, making strict compliance with Section 6(b)(5) impractical for crypto exchanges. The traditional regulatory framework assumes a level of control over trading venues that simply does not exist in crypto markets.

3. Regulatory Guardrails for Centralized and Decentralized Exchanges

This paper is not intended as an indictment of the SEC's crypto policy. We recognize that the agency's actions have been driven by a genuine effort to protect American investors and foster trust in crypto exchanges. Our criticism is instead directed at the legislative branch, which has long failed to establish a clear regulatory framework for the crypto industry. In the absence of congressional action, the SEC had little choice but to apply the existing regulatory system to crypto markets. However, U.S

President Donald Trump has signaled a potential shift in this approach. Following Gary Gensler's resignation as SEC Chair, Trump appointed a successor more aligned with the crypto industry's interests, marking a departure from the prior administration's enforcement-heavy strategy (SEC 2025). As a result, the SEC has paused or dropped enforcement actions against major actors such as Coinbase and the developers of the Uniswap protocol. Looking ahead, President Trump has announced his intention to introduce legislation that would establish a clear, tailored regulatory framework for digital asset markets (Executive Order No. 14099, Jan. 23, 2025).

With this in mind, we aim to contribute to the ongoing discussion by proposing broad regulatory guardrails for both centralized and decentralized exchanges. Regulation that incorporates these guardrails is more likely to promote capital formation by enhancing legal certainty, while also strengthening investor protection. Tailored legislation that aligns with these overarching policy goals, can play a critical role in fostering the trustworthiness of crypto exchanges and the broader digital asset ecosystem.

First, crypto regulation in the U.S. should begin by resolving the ongoing debate over whether crypto assets are securities under the Howey Test. The Howey Test determines whether a transaction qualifies as an investment contract – and thus a security (Rosenberg 2020). Legislation should contain a technology-neutral definition of crypto assets and assign oversight of the entire crypto market to a single federal agency.

Second, the regulation of centralized exchanges must account for the fact that their business model fundamentally differs from that of traditional securities exchanges. Coinbase and other centralized crypto exchanges operate more like broker-dealer internalizers – they custody assets and manage a limit order book for their customers rather than merely facilitating trades between independent parties. Regulatory frameworks must reflect this distinction by prioritizing custody requirements and the safeguarding of customer funds. Centralized exchanges should be required to verify that they hold customer assets 1:1.

Third, sunlight is still the best disinfectant. The famous quote by U.S. Supreme Court Justice Louis Brandeis (Brandeis 1913), alluding to the value of disclosure as a regulatory instrument in securities markets, still holds true today. However, the current disclosure regime under the 1934 Act, focused on issuers providing disclosures, is ill-suited for crypto assets. Disclosure requirements for crypto assets need to be rethought. The decentralized and open-source nature of many crypto networks eliminates the traditional information asymmetry that underpins investing in corporate securities. For some crypto assets, disclosure may not be necessary at all. For others, where identifiable individuals or entities exert significant influence over the project, tailored disclosure obligations could be appropriate. In certain cases, exchanges themselves could be required to provide relevant disclosures about listed assets (Coinbase 2021).

Fourth, rather than attempting to restrict the development of decentralized exchanges, regulation should focus on the applications that enable investors to interact with these protocols (Jennings 2022).

Fifth, the liquidity of decentralized exchanges can facilitate the regulatory process. While decentralized protocols can be copied and redeployed to circumvent regulation, liquidity remains the defining competitive advantage. The securities market adage, “Liquidity begets liquidity,” applies to decentralized exchanges as well – established decentralized exchanges dominate because they offer deeper liquidity.

Sixth, developers who retain meaningful control over decentralized protocols, should be subject to regulatory oversight. The 2020 Uniswap-SushiSwap episode highlights that protocol developers often retain more control over deployed protocols than expected. SushiSwap, a copied version of the Uniswap protocol, attempted to lure away Uniswap's liquidity by offering more attractive incentives to liquidity providers. In response, Uniswap implemented a reward system, distributing it strategically to liquidity providers and users (Gemini n.d.-b). This successful defense demonstrated that protocol developers can – in certain cases – influence protocols even after deployment.

Seventh, regulators should keep in mind that even when developers relinquish control over decentralized protocols, alternative avenues for fostering trust via regulation remain available. Decentralized exchanges often issue governance tokens, which enable network participants to take part in protocol decisions, for example, by allowing token holders to vote on proposed changes such as transaction fees or protocol upgrades. From a regulatory perspective seeking to build trust by focusing on identifying clearly accountable actors, this may seem counterproductive. The approach strengthens participatory structures, but also means that developers no longer retain meaningful control. This potentially removes them as the primary targets of regulatory accountability. However, governance tokens offer an alternative model for building trust – one grounded in the capability approach (cf. Kaminski, Düwell and Richter 2024). By distributing governance rights to token holders, the system enhances option values (by providing users with concrete means to influence the protocol) and legacy values (by enabling influence over the protocol's long-term development). Governance tokens thus reflect a shift in responsibility in decentralized systems from a centralized authority to the structured distribution of agency to all users. This in itself may serve as an alternative focal point for regulatory engagement.

Finally, the federal supervisory agency should be responsible for establishing auditing standards for decentralized protocols. Decentralized exchanges could be evaluated based on security and other key factors, similar to how credit ratings are assigned in traditional financial regulation (Armour et al. 2016: 127). The agency should assess and grade protocols that reach certain thresholds indicating widespread use by the American public. Applications enabling users to interact with these protocols should be required to display the assigned ratings, ensuring that users are informed of the protocol's security and reliability before engaging with it.

References

Adams, Hayden (2024): X post, 10 April, 3:45 PM, https://x.com/haydenzadams/status/1778126466984575166, last access: February 2, 2025.

Armour, John et al. (2016): Principles of Financial Regulation, Oxford: Oxford University Press.

Beny, Laura N. (2002): "U.S. Secondary Stock Markets. A Survey of Current Regulatory and Structural Issues and a Reform Proposal to Enhance Competition", in: Columbia Business Law Review, 2002(2), pp. 415–508.

Brandeis, Louis D. (1913): "What Publicity Can Do", in: Harper's Weekly, December 31, 1913, pp.×10–13, https://www.sechistorical.org/collection/papers/1910/1913_12_20_What_Publicity_Ca.pdf, last access: May 1, 2025.

Coffee, John C., Sale, Hillary A. and Whitehead, Charles K. (2020): Securities Regulation. Cases and Materials, 14th ed., St. Paul, MN: Foundation Press.

Coinbase (2021a): "Digital Asset Policy Proposal. Safeguarding America's Financial Leadership", October 24, 2021, https://www.coinbase.com/en-de/blog/digital-asset-policy-proposal-safeguarding-americas-financial-leadership, last access: February 2, 2025.

Coinbase (2021b): Registration Statement (Form F-1), 12, February 25, 2021, https://www.sec.gov/Archives/edgar/data/1679788/000162828021003168/coinbaseglobalincs-1.htm, last access: February 2, 2025.

Coinbase (n.d.-a): "Liquidity Program Overview", https://www.coinbase.com/exchange/liquidity-program last access: February 2, 2025.

Coinbase (n.d.-b): "Trading Rules", https://www.coinbase.com/en-de/legal/trading_rules, last access: February 2, 2025.

Coinbase (n.d.-c): "What Does Coinbase Do With My Digital Assets?", https://help.coinbase.com/en/coinbase/other-topics/legal-policies/what-does-coinbase-do-with-my-digital-assets, last access: February 2, 2025.

Committee on Capital Markets Regulation (2023): "Crypto Exchanges Cannot Register with the Securities and Exchange Commission", June 6, 2023, https://capmktsreg.org/wp-content/uploads/2023/06/CCMR-Crypto-Exchanges-Cannot-Register-With-the-SEC-06-06-23.pdf, last access: February 3, 2025.

Dombalagian, Onnig H. (2024): "Serendipity and Self-Regulation. The Evolution of Cryptocurrency-Based ETPs", Tulane Public Law Research Paper No. 24–9.

Fleckner, Andreas M. (2006): "Stock Exchanges at the Crossroads", in: Fordham Law Review 74, pp. 2251–2303.

Fox, Merritt B. et al. (2018): Securities Market Issues for the 21st Century, New York: Columbia Law School.

Ge Huang, Vicky, Osipovich, Alexander and Kowsmann, Patricia (2022): "FTX Tapped Into Customer Accounts to Fund Risky Bets, Setting Up Its Downfall", in: The Wall Street Journal, November 10, 2022, https://www.wsj.com/articles/ftx-tapped-into-customer-accounts-to-fund-risky-bets-setting-up-its-downfall-11668093732, last access: February 2, 2025.

Gensler, Gary (2024): "Statement on Spot Bitcoin ETFs", U.S. Securities and Exchange Commission, January 10, 2024, https://www.sec.gov/newsroom/speeches-statements/gensler-statement-spot-bitcoin-011023, last access: February 3, 2025.

Gemini Cryptopedia (n.d.-a): "Decentralized Finance and Impermanent Loss", https://www.gemini.com/cryptopedia/decentralized-finance-impermanent-loss-defi, last access: February 2, 2025.

Gemini Cryptopedia (n.d.-b): "SushiSwap, Uniswap, and the Vampire Attack", https://www.gemini.com/cryptopedia/sushiswap-uniswap-vampire-attack, last access: February 3, 2025.

Harris, Larry (2002): Trading and Exchanges. Market Microstructure for Practitioners, Oxford: Oxford University Press.

Jennings, Miles (2022): "Regulate Web3 Apps, Not Protocols", a16z, September 29, 2022, https://a16zcrypto.com/posts/article/web3-regulation-apps-not-protocols/, last access: February 2, 2025.

Johnson, Kristin N. (2023): "Regulating Cryptocurrency Secondary Market Trading Platforms", in: University of Chicago Law Review Online, https://lawreview.uchicago.edu/online-archive/regulating-cryptocurrency-secondary-market-trading-platforms#heading-8, last access: February 2, 2025.

La Porta, Rafael et al. (1997): "Legal Determinants of External Finance", in: The Journal of Finance 52(3), pp. 1131–1150.

Kazimirov, Alexandros (2024): "Regulation by Enforcement. Why the Securities and Exchange Commission's Vision for Digital Asset Markets Lacks Clarity and an Alternative European Model", Stanford-Vienna Transatlantic Technology Law Forum (TTLF) Working Paper No. 118, https://law.stanford.edu/publications/no-118-regulation-by-enforcement-why-the-securities-and-exchange-commissions-vision-for-digital-asset-markets-lacks-clarity-and-an-alternative-european-model/, last access: February 2, 2025.

Koome, Brian (2021): "Uniswap V1 Whitepaper. How the User-Centric Protocol Empowers Liquidity Providers", Cryptopolitan, March 12, 2021, https://www.cryptopolitan.com/uniswap-v1-whitepaper-how-the-user-centric, last access: February 2, 2025.

Lee, Larissa (2016): "New Kids on the Blockchain. How Bitcoin's Technology Could Reinvent the Stock Market", in: Hastings Business Law Journal 12, pp. 122–158.

Lin, Lindsay X. (2019): "Deconstructing Decentralized Exchanges", in: Stanford Journal of Blockchain Law & Policy 2, pp. 62–91.

Lutz, Sander (2022): "SEC Chair Gensler Threatens Action Against Unregistered Crypto Exchanges", Decrypt, May 18, 2022, https://decrypt.co/100806/sec-chair-gensler-threatens-action-against-unregistered-crypto-exchanges, last access: February 1, 2025.

(NYSE) New York Stock Exchange (n.d.): "History of the NYSE", https://www.nyse.com/history-of-nyse, last access: February 2, 2025.

Rosenberg, Peter (2020): "When They Howey, We All Howey", in: Fordham Journal of Corporate & Financial Law Blog, January 5, 2020, https://news.law.fordham.edu/jcfl/2020/01/05/when-they-howey-we-all-howey/, last access: February 2, 2025.

Soderquist, Larry D. and Gabaldon, Theresa A. (2014): Securities Law. Concepts and Insights, 6th ed., New York: Foundation Press.

Shahin, Mohammad (2020): "Understanding Smart Contracts. The Vending Machine Analogy", Medium, October 1, 2020, https://shahinms.medium.com/understanding-smart-contracts-the-vending-machine-analogy-ce7a4cd74fb3, last access: February 2, 2025.

Uniswap (n.d.): "Wells Submission on Behalf of Uniswap Labs", https://blog.uniswap.org/the-fight-for-defi-continues, last access: February 2, 2025.

Uniswap (n.d.): "What is a Liquidity Pool?", https://support.uniswap.org/hc/en-us/articles/8829880740109-What-is-a-liquidity-pool, last access: February 2, 2025.

Uniswap (n.d.): "What is Price Impact?", https://support.uniswap.org/hc/en-us/articles/8671539602317-What-is-Price-Impact, last access: February 2, 2025.

Westbrook, Tom (2024): "Global Crypto Market Tops $3 trillion on Hopes of Trump-fuelled Boom", Reuters, November 14, 2024, https://www.reuters.com/technology/crypto-market-capitalisation-hits-record-32-trillion-coingecko-says-2024-11-14/, last access: February 2, 2025.

Legal Sources

Executive Order No. 14099, January 23, 2025, Strengthening American Leadership in Digital Financial Technology, 90 Fed. Reg. 1.

(SEC) U.S. Securities and Exchange Commission (2025): "SEC Crypto 2.0. Acting Chairman Uyeda Announces Formation of New Crypto Task Force", Press Release No. 2025–30, January 21, 2025, https://www.sec.gov/newsroom/press-releases/2025-30, last access: February 2, 2025.

(SEC) Securities and Exchange Commission v. Coinbase, Inc., No. 23-cv-06438 (S.D.N.Y. July 26, 2023), https://www.sec.gov/files/litigation/complaints/2023/comp-pr2023-138.pdf, last access: February 2, 2025.

(SEC) U.S. Securities and Exchange Commission (2023), Litigation Release No. 25751, June 21, 2023, https://www.sec.gov/litigation/litreleases/2023/lr25751.htm, last access: February 2, 2025.

(SEC) U.S. Securities and Exchange Commission (2021), Custody of Digital Asset Securities by Special Purpose Broker-Dealers, Release No. 34–90788, April 27, 2021, https://www.sec.gov/files/rules/policy/2020/34-90788.pdf, last access: February 3, 2025.

(SEC) U.S. Securities and Exchange Commission (2018), Release No. 34–83723, File No. SR-BatsBZX-2016-30, May 6, 2018.

Supplemental Information and Reopening of Comment Period for Amendments Regarding the Definition of "Exchange", 88 Fed. Reg. 29,448 (May 5, 2023).

III. Enabling Conditions for Trust and Responsibility

Ethics and Regulation of AI Systems in Medicine

The Example of Cancer Detection

Sebastian Bartsch, Marcus Düwell, Jan-Hendrik Schmidt, Alexander Benlian

Abstract *The integration of artificial intelligence (AI) systems into medical practice, specifically in cancer detection, presents unknown opportunities for better diagnoses and treatments for patients. However, with the integration of AI systems into a traditional relationship between healthcare professionals and patients, questions regarding accountability in this expanded relationship arise since traditional standards of medical law and medical ethics are addressed towards a healthcare professional. Against this backdrop, we investigate the necessary capacities to hold each involved party accountable (i.e., the healthcare professional, the patient, the developer, a regulatory oversight, and perhaps a clinical AI expert to support the healthcare professional). For this, we first explore the ethical and regulatory implications of employing AI systems in healthcare. We stress that the possibility of maintaining accountability is of central importance for the acceptability of the implementation of AI systems. As AI systems are often inscrutable and do not allow any party to explain and justify the behavior of the AI system, we examine whether and how explainable AI (XAI) methods can support each party with their accountability obligations. With our considerations, we propose a theoretical model for distributing accountability among each involved party and finally highlight the need for regulatory frameworks that can enable an ethically acceptable development and use of AI systems.*

1. Introduction

As the performance of AI systems continues to advance, they increasingly surpass human capacities in various domains. The more this will be the case, the more questions can be raised about the responsible development, deployment, and use of AI systems in practice. Given the variety of contexts in which AI systems are applied, each with its unique implications and challenges, we will limit our focus to the use of AI systems in medicine. In these contexts, the specific relationship between healthcare professionals and patients is of central importance, and therefore, the introduction of AI systems raises particular questions. For this reason, we also exclude possible AI systems designed for direct interaction with persons outside the regulated healthcare environment. These systems would introduce distinct ethical questions regarding moral and legal responsibilities outside the highly regulated context of the healthcare system. In case AI systems take over tasks traditionally held by healthcare professionals, the challenge is that healthcare professionals do not only provide medical advice, but some sophisticated hermeneutical skills are required. By this we mean the following: Often patients are not fully able to understand the complicated medical information and they are often not aware of how they can respond to them. The healthcare professional needs skills to translate the information into a language a patient can understand and to interpret the wishes, expectations, hopes, and fears of the patient. Such hermeneutical skills are a necessary precondition for an accountable advice by the healthcare professional. This becomes more complicated if AIs are introduced in this relationship, and it raises complex ethical questions regarding standards of medical ethics that have been debated for decades (e.g., Fiske et al. 2019; Bringsjord 2008).

In this article, we focus on AI systems in the context of cancer detection. Focusing on this context is of high interest, as these AI systems are able to incorporate various data inputs to enable healthcare professionals to make more advanced and accurate diagnoses and treatments. However, our chosen example clearly presupposes that AI systems are solely used by the healthcare professional (perhaps with support from a clinical AI expert) and that AI systems are not directly used or interact with parties outside this healthcare setting. Additionally, we focus on such specific AI systems, as the example we have chosen is discussed in the overall context of this volume, focusing on normative questions that can be asked at a reflective level. Thus, we are not primarily evaluating whether some developments in AI systems are, as such, desirable, good, or risky. We rather focus on a question that necessarily has to be discussed before such an evaluation is possible, namely the question of what kind of capacities are required from the healthcare professionals as users of an AI system and how the relationship between the design of an AI system and the potential capacities of the healthcare professionals can raise questions of accountability. We are particularly interested in the question of how and to what extent healthcare professionals are capable of making informed judgments based on the advice AI systems provide. Suppose healthcare professionals are unable to understand how advice from AI systems is generated, and this raises the question to what extent the healthcare professional could be held accountable for the diagnosis and the treatment that is initiated on the basis of AI systems' advice. As a follow-up question, we will ask whether possible problems of accountability can be overcome by introducing clinical AI experts and oversight committees. Finally, we assess whether technical solutions like XAI are sufficiently capable of supporting healthcare professionals in their ability to deal with those systems in a responsible way.

We begin our article by (1) describing the technical functionality of AI systems in cancer detection, followed by (2) an exploration of the normative concerns these systems raise. We then (3) evaluate the capacity of XAI to address the raised concerns and (4) discuss the need for additional regulatory measures to respond to those concerns.

2. AI Systems for Cancer Detection – How Does it Function?

Recent advancements in AI systems have shown significant promise in enhancing the accuracy and efficiency of cancer detection and diagnosis by supporting healthcare professionals (Zheng et al. 2023; Nassif et al. 2022). For instance, AI systems are capable of detecting various types of cancer, including skin cancer (e.g., Brancaccio et al. 2024), prostate cancer (e.g., Perincheri et al. 2021), gastrointestinal cancer (e.g., Suzuki et al. 2021), lung cancer (e.g., Ponnada and Srinivasu 2019), and, in particular, breast cancer (Zheng et al. 2023; Nassif et al. 2022). In general, such AI systems require cross-sectional images of suspected cancerous areas. In the example of detecting breast cancer, those images are often acquired through imaging modalities such as mammography for early-stage breast cancer detection (Zheng et al. 2023). Once captured, these images are processed and analyzed by the AI system, which has been trained to identify specific cancer types with high precision (Sechopoulos et al. 2021). Thus, the AI system is not omniscient and is only able to detect one type of cancer on which the AI system is trained and optimized (e.g., Russell and Norvig 2016).

These limitations of AI systems usually stem from the design and training process of the AI system (e.g., Russell and Norvig 2016). AI systems for cancer detection often employ a supervised learning approach, where the AI system is designed and trained for a single, specific diagnostic purpose (e.g., Shravya et al. 2019; Osareh and Shadgar 2010; Gupta and Gupta 2018). In this supervised learning training approach, developers train the AI system on a labelled dataset comprising both positive examples (i.e., images that contain the cancer to be identified) and negative examples (i.e., images that do not contain the cancer to be identified; e.g., Cunningham et al. 2008; Beeravolu et al. 2021). Accordingly, this training approach assigns the AI system a well-defined task and their desired solution of the task (often described as ground truth labels), enabling the AI system to iteratively adjust its parameters using statistical techniques until it generalizes the problem of detecting cancer with high accuracy. This iterative refinement process, referred to as *learning*, allows the AI system to progressively improve its predictive capabilities for the focused cancer type (Cunningham et al. 2008).

In designing AI systems for cancer detection, developers often employ neural networks that emulate the functioning of the human brain (Sechopoulos et al. 2021; Krogh 2008). A typical neural network comprises an input layer, one or more hidden layers, and an output layer, each containing nodes that perform specific mathematical operations when activated (Krogh 2008). These nodes on each layer are interconnected to form the network architecture, and developers can adjust the network's complexity by modifying the number of hidden layers and nodes to address various problems (Krogh 2008; Russell and Norvig 2016). Complex problems often require deeper neural networks with numerous layers and nodes, while simpler neural networks may suffice for less demanding problems (e.g., Hunter et al. 2012). However, as neural networks become more complex, they increasingly appear as *black boxes*, making it challenging for developers and users to understand how their AI systems derive specific outcomes (Asatiani et al. 2020; Berente et al. 2021). Through this resulting lack of transparency, developers may still evaluate an AI system's performance on test images by comparing the amount of correct or incorrect outcomes. However, they often cannot fully explain the reasoning behind the outcomes and are, thus, unable to fully understand the decision-making process of the AI system.

Since this lack of transparency is often critical in the productive use of AI systems, research on XAI seeks to address the challenge of understanding AI systems despite their inherent complexity (e.g., Adadi and Berrada 2018; Dwivedi et al. 2023). Notable advancements in XAI have led to the development of methods that analyze AI systems' behavior through external observation, enabling developers to interpret the system's functioning without direct insight into its internal architecture (e.g., Friedman 2001; Apley and Zhu 2020). By observing the AI systems' behavior, developers can derive aspects of the AI system's internal decision-making processes. Thus, developers approximate the AI systems' internal decision-making process but do not get a detailed explanation due to the limited interpretability of the system's inner architecture and mechanisms (Molnar 2019). Nevertheless, XAI methods provide developers transparency into their AI systems, with the aim of fostering user trust and acceptance of their AI systems (e.g., Dwivedi et al. 2023; Shin 2021). In this context, XAI methods can generally be divided into post-hoc and ante-hoc methods to gain transparency into the behavior of an AI system (Retzlaff et al. 2024).

Ante-hoc methods provide explanations inherently based on their algorithmic design and, thus, are limited to AI systems that use transparent algorithms such as linear regression and simple decision trees (Molnar 2019). On the other hand, post-hoc models aim to explain any designed and implemented AI systems by constructing additional surrounding models to explain either the AI system as a whole or specific instances within the dataset (Retzlaff et al. 2024). However, relying on a surrounding model implies that only an approximation of the AI system is explained, rather than the AI system itself. Consequently, post-hoc explanations may be inaccurate due to the inherent limitations of approximations, potentially undermining trust in their validity (Rudin 2019). Consequently, while ante-hoc models are generally more aligned with accurate explanations, this does not automatically mean that post-hoc approaches are inherently incapable of providing meaningful insights into the behavior of an AI system (e.g., Linardatos et al. 2020; Molnar 2019).

Classical representatives of post-hoc methods are partial dependency plots (PDP), permutation feature importance (PFI), local interpretable model-agnostic explanations (LIME), and influential instances (Molnar 2019). These representatives can further be divided into global-agnostic models, local-agnostic models, and instance-based methods to gain transparency into AI systems at different stages of the AI lifecycle (ibid.). In particular, global-agnostic models explain the AI system as a whole, while local-agnostic models explain the individual decisions of an AI system (Hariharan et al. 2023). Lastly, instance-based methods are used within the development of AI systems to gain information on how the AI system learns from the given dataset (Molnar 2019). Table 1 contains an overview of frequently used XAI methods and provides an explanation of each XAI method.

Table 1: Overview of frequently used XAI methods.

Category	XAI Method (Examples)	Explanation of the XAI Method	Applicability
Global-agnostic	Partial Dependency Plots (PDP)	PDPs illustrate the marginal effect of a feature (e.g., age, gender, or body weight) on the prediction of the AI system by averaging over all other features. Thus, PDP visualizes how changes in a given feature influence the prediction of the AI system. For example, a PDP can show how the probability of getting cancer changes when age increases by one year (Friedman 2001; Goldstein et al. 2015).	PDP is useful for understanding the importance of features during AI systems' development and validation.
Global-agnostic	Permutation Feature Importance (PFI)	PFI follows a similar principle compared to PDP but instead assesses feature importance by randomly shuffling values of a specific feature and measuring the resulting change in model performance. Thus, PFI quantifies how much a feature contributes to the prediction of the AI system. For example, while PDP reveals how the probability of getting cancer changes with age, PFI visualizes the relative importance of age for the prediction compared to other features, such as gender or body weight (Breiman 2001; Fisher et al. 2019).	PFI can be used to understand the dependency of the features of making predictions and can be used to validate and debug the AI system.

Category	XAI Method (Examples)	Explanation of the XAI Method	Applicability
Local-agnostic	Local Interpretable Model-agnostic Explanations (LIME)	LIME generates synthetic input data similar to a given instance and trains a transparent, interpretable AI system (e.g., linear regression) to approximate local decision boundaries. These boundaries provide users with insights into how the AI system behaves at a specific point. However, LIME's explanations are strictly local, meaning they apply only to a single instance rather than the entire AI system. For example, LIME can explain why an AI system predicts that a 58-year-old man weighing 65 kg has a low probability of developing cancer by showing that a normal body mass index decreases the risk of getting cancer. However, LIME cannot generally explain how the features of age, year, and body weight interact with each other and affect cancer (Ribeiro et al. 2016).	LIME can explain individual predictions to users after deploying the AI system.
Instance-based	Influential Instances	Influential instances are data points within the training set that significantly impact the predictions of the AI system. These instances can be problematic as their removal may cause substantial changes in AI systems' behavior. Identifying and managing influential instances enhances AI systems' robustness and performance while providing valuable insights into dataset composition (Molnar 2019).	Influential instances are useful during data preprocessing and debugging AI systems, as well as understanding the dataset.

Applying XAI methods in the context of cancer detection, healthcare professionals are not only informed of the presence and location of potential cancerous regions in the provided image but also receive insights into the AI system's decision-making process (Tosun et al. 2020). This additional information aims to enable healthcare professionals to integrate both the diagnostic findings and the AI systems' interpretative behavior into the patient's anamneses, guiding subsequent clinical decisions. Importantly, for these types of AI systems, even if they are used to support healthcare professionals with their clinical decisions, healthcare professionals remain the primary decision-makers, retaining the authority to accept or override the suggestions of the AI system (e.g., Tosun et al. 2020). Thus, this non-autonomous design of the AI systems aims to ensure that the system serves as a decision support system, with ultimate decision-making power residing with the healthcare professionals, who are accountable for determining the course of treatment (Cooper et al. 2022). This specific setting leads, however, to the question of whether the healthcare professional is able to deal with these AI systems in a responsible way – a question that shall now be addressed.

3. Ethical Questions

In the first instance, it seems evident that the healthcare professional is still in charge of the situation. The healthcare professional bears accountability vis-à-vis the patient for diagnosing and treating the patient, with the AI system merely providing suggestions. Ultimately, healthcare professionals make the final decision, implying that their decisions and resulting actions can still be evaluated based on established standards of medical ethics and medical law. From this perspective, there seems to be no immediate need for additional regulatory adjustments, as existing frameworks primarily regulate the actions of healthcare professionals (e.g., Gilvary et al. 2019; De Boer and Kudina 2021; Kudina and de Boer 2021; Jiang et al. 2017; Lee 2022; Morley et al. 2020; Schleidgen and Friedrich 2023; Rudschies and Schneider 2024; Speck et al. 2021).

However, as AI systems become more advanced and more integrated into the traditional relationship between healthcare professionals and patients, the consideration of AI systems within this context becomes more complex. In this context, it is quite likely that, to some extent, healthcare professionals will not be able to fully understand the AI system and, in particular, how the AI system arrives at its diagnoses and treatment proposals. This is, in some sense, even the goal and the rationale behind the use of the technology: AI systems are designed to enhance diagnostic and treatment capabilities beyond what healthcare professionals alone can achieve – if AI systems would not offer such superior analytical capacities, their use in the medical context would be questionable right from the beginning. However, there are still important reasons to assume that the capacities of AI systems are so impressive that it would be prima facie irresponsible not to use them in light of the possible positive impacts on the well-being of the patient. However, the more AI systems are capable of fulfilling their task successfully, the more they will disrupt the traditional interaction between healthcare professionals and their patients.

This disruption creates an inherent tension: If AI systems were ineffective in cancer diagnoses, it would be pointless to use them. However, the more proficient they become, the more likely it is that their capacities exceed the interpretative capacities of healthcare professionals. We have to keep in mind here that cancer treatment is not only a technical process where we have to evaluate biological processes and the probabilities of successful medical interferences. Rather, it is a process where healthcare professionals must deal with particular, individual situations of patients, their wishes, fears, hopes, and the outlook of life of their patients. Whatever AI systems suggest, we must expect that the healthcare professional is still in the position to mediate between the available medical information and the specific situation of their patients, and the interference of AI systems should support them in their tasks to arrive at diagnoses and treatments. The integration of AI systems raises at least the following two questions.

First, what do we have to presuppose about the capacities of healthcare professionals regarding the use of AI systems? To what extent are these persons capable of understanding the way AI systems work? While healthcare professionals are not expected to be experts in AI systems, they must still be capable of assessing the reliability and implications of AI-generated recommendations to evaluate the value of the retrieved diagnostic and treatment proposals. Looking ahead, as AI systems become even more sophisticated, we can ask: At what point does it become unreasonable to expect healthcare professionals to be capable of understanding why an AI system has made a particular suggestion? Of course, we cannot specify the technical details that healthcare professionals need to know. However, suppose we assume that there is a continuum between a simple layperson who is just able to use a computer and an IT expert. In that case, we should wonder at which point we would locate the healthcare professional in order to ask: Is this person still fully in charge of the diagnostic and treatment process?

Second, what do we have to presuppose about the qualities of an AI system that we would ask: This AI system is designed in a way that healthcare professionals with mediocre technical knowledge are still sufficiently capable of understanding why an AI system retrieved a specific proposal? And what kind of adjustments to AI systems are sufficient to enable the healthcare professional to have a sufficient understanding?

These questions are critical because, as long as healthcare professionals are responsible for treatment decisions, they must also remain accountable for them. Since we deal with cancer detection, it may easily be a question of life and death. This implies that healthcare professionals are fully accountable for their decisions; to emphasize the importance of this point: If something goes wrong, their decisions and actions could be assessed in the courtroom. For the understanding of healthcare professionals, we have to make realistic assumptions about the technical knowledge of a mediocre person who received training in medicine, and not AI systems. Of course, we can assume that future healthcare professionals will have more advanced technical knowledge compared to today, but it would be unrealistic to assume that healthcare professionals will be IT experts. This leads to further questions: How much expertise should be required to hold healthcare professionals accountable for their decisions? And how should an AI system be designed so that it is capable of providing the healthcare professional with the information needed to make those decisions responsibly?

Suppose it were the case that the proposals of the AI system significantly exceed the understanding of healthcare professionals. In that case, it may be necessary to rethink the division of labor: Thus, we can think about a laboratory for cancer detection in which healthcare professionals are still in charge of their medical decisions and communication with their patients, but a clinical AI expert would be accountable for commanding the AI system and the interpretation of the suggestions of the AI system. In such cases, we come to the same questions regarding the necessary features of the AI system and the required capacities of the clinical AI experts, but we would additionally have to ask: What would be required regarding the capacities of the healthcare professionals and the clinical AI expert to communicate successfully with each other? And if they make a mistake, who will be held accountable in the courtroom? Can we determine under what conditions healthcare professionals will be in the dock, and under what conditions the clinical AI expert will be, and when both might be held accountable? If we were not capable of determining those questions on the level of general criteria, we would have a typical case of accountability diffusion, often discussed under the label 'responsibility gap' (see, e.g., Düwell 2012, pp. 185–192; Matthias 2004), thus, situations where nobody is accountable. In a situation where the life and death of persons are at stake that is unacceptable.

If we have difficulties determining those things, we could further ask whether there is a possibility to enhance the feedback procedures regarding the functioning of the AI system and whether we could install some form of general oversight of the process. Both things are necessary, anyhow, since even if accountability frameworks are well-defined at one point in time, ongoing technological advancements will necessitate continual reassessment. Thus, what are the requirements regarding the feedback procedures where the developer of the AI system (let us call them 'D') to ensure that the AI system is designed appropriately to enable healthcare professionals with mediocre technical abilities (let's call them 'H'), perhaps with the support of clinical AI experts (let us call them 'E'), to be able to evaluate the suggestions of the AI system? Additionally, what are the requirements regarding institutions that are capable of exercising oversight (let us call them 'O')? O must be in a position to judge whether the relationship between the features of the AI system and those capacities that one can reasonably expect from healthcare professionals (plus clinical AI experts) to have an appropriate accountability relationship towards the patient (let us call them 'P').

In the scheme below, we illustrate our proposed model by focusing on accountability relationships (see Figure 1). It highlights that healthcare professionals remain ultimately accountable to their patients. In some cases, the healthcare professionals may require technical assistance from the clinical AI expert (let us call them 'E'), while in other cases, they may independently interpret the suggestions of the AI system. Developer D is only accountable for realizing the AI system in ways that are required to bring healthcare professionals H in a position to realize accountability vis-à-vis patient P. The regulatory authority O is accountable for the oversight that the use of AI systems is regulated in a way that healthcare professionals H can fulfill their tasks vis-à-vis their patients P. In this vein, it should be emphasized that these considerations are fully independent of the content of the normative standards that are applied and the specific ethical theory that is presupposed (for an overview of possible ethical approaches, see Düwell 2012). Whatever normative standards one applies, it always has to be presupposed that healthcare professionals H are capable of applying them in a specific practice. That implies that they possess the capacity to understand the practice to the extent that they are able to interfere according to those standards.

Figure 1: Proposed model for accountability relationships

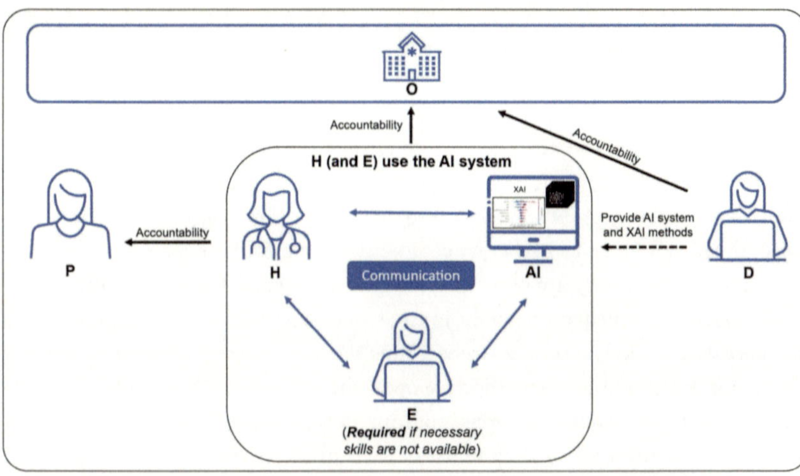

In summary, we can systematize the relevant capacities and accountability relationships:

(1) H must be capable of understanding the proposals of the AI systems, potentially with the support of E, and he or she is accountable towards P, and in case of failure of H, P must have legal possibilities to address this.
(2) E is only responsible for H. E must have the capacity to deal with the AI system competently and understand the questions of H.
(3) O must be capable of providing feedback to D regarding the appropriate functioning of AI systems. D is accountable to O. Of course, there can and should be direct communication from H and E with D, but the line of accountability goes towards O.

(4) H (and E) are accountable to O. O must be capable of assessing whether AI systems are designed in a manner that enables H to take accountability for medical decisions vis-à-vis the patient.

We have to assume that it is, in principle, possible that H, E, D, and O can end up in a courtroom because of a failure regarding their part of the accountability. Since we talk about serious decisions (decisions about life and death), it must be possible to differentiate between the respective parts of the accountability. It is particularly important to stress that the AI system must be designed, developed, and deployed in a way that the patient has the possibility to claim his or her rights in the courtroom. The only instance against which the patient can claim rights is the healthcare professional. If that were not the case, patient rights would be seriously endangered. This, however, brings the healthcare professional into a much more vulnerable position. Therefore, it is necessary to specify what healthcare professionals can be held accountable for and where this accountability ends. Without such differentiation, accountability could either be entirely absent or unfairly distributed among all parties, both of which are unacceptable outcomes. The question for the following part is, therefore, what are the possibilities from the perspective of XAI to provide us with criteria for the differentiation of the respective parts of the accountability?

4. Possible Answers to these Challenges From the XAI Perspective

In what follows, the role of XAI methods is discussed in enhancing transparency at different stages of the design, development, and use of AI systems (e.g., Molnar 2019). These methods aim to assist each actor involved in fulfilling their unique obligations by providing insights that enable explanations and justifications of decisions to their respective counterparts (e.g., the patient P or regulatory authorities O). Below, we analyze each accountability relationship presented in the previous section and discuss how XAI methods can enhance the transparency of the AI system and support the involved parties with their accountability obligations.

D is accountable to O. When developing medical AI systems, we suggest that D should not be allowed to design and develop any AI system without requirements provided by the regulatory authorities O. Thus, D should be held accountable to meet the requirements of O, as specified in how medical AI systems can be designed and developed. In particular, such requirements should include accountability obligations requiring explanation and justification about the performance of the AI system and requiring explanation and justification about the functionality of their AI systems to communicate the limits and constraints under which the AI system can be used with minimal risk of ethical issues. To support D with these accountability obligations, they need to ensure high performance (i.e., high efficiency with minimal error rates) of the AI system and need to understand the behavior of their AI system. This implies that D needs transparency about the training process to ensure high performance and needs transparency about the behavior of their AI systems. In this context, D can rely on XAI methods that help them to understand and debug their training data, such as influential instances, which allow D to increase the performance of their AI system. Additionally, D can rely on global model-agnostic methods, such as partial dependency plots, to comprehend the behavior of their AI system, which allows D to communicate the limits and constraints of their AI systems.

XAI methods, such as influential instance analysis (see Table 1), allow D to evaluate the impact of specific data points on the performance of their AI system (Molnar 2019). By identifying and removing specific data points that increase the error rate of their AI system, D can better understand how their AI system learns from the given training data. This allows D to enhance their training data and increase the subsequent performance of their AI system, thereby minimizing harmful outcomes such as incorrect diagnoses and medical treatments. As a result, this process supports D in explaining and justifying the robustness of their AI system to O and fulfills their accountability obligations.

Additionally, XAI methods like partial dependence plots (Friedman 2001) and permutation feature importance (Fisher et al. 2019) provide D with a high-level understanding of the behavior of their AI system (Molnar 2019). This high-level understanding of their AI system enables D to ensure that the AI system operates as intended and to detect unintended biases. This information allows D to edit parameters or change the architecture of their AI system to ensure intended behavior. Importantly, the abstract level of explanation provided by these XAI methods does not require D to have in-depth domain knowledge of medicine, making it practical for developers to verify the behavior of their AI system and communicate its limits and constraints to O.

H (and E) are accountable to O. We suggest that O acts as the intermediary between D and H (and E), holds D accountable for overseeing the performance of the AI system, and signs the suitability of the AI system for deployment. To fulfill this accountability obligation and oversee the AI system, O needs experts in both fields (i.e., medicine and AI development) to evaluate the performance from a technical and semantic level. For the evaluation, O needs to rely on global model-agnostic methods provided by D to understand and evaluate the behavior of the AI system. These XAI methods help O check compliance with specified requirements by avoiding or mitigating potential ethical issues. Based on this assessment, O can approve or reject the AI system and provide D with additional information on how to improve their AI system.

Once O scrutinized the AI system, we suggest that H (and E) should be accountable for understanding the limits and constraints of the AI system. This ensures that

H (and E) are aware that the AI system might suggest false diagnoses or medical treatments, and that they should not blindly follow the suggestion of the AI system.

H is accountable to P. H bears the accountability for the diagnosis and medical treatment of their patients. Although H might interact with the AI system and draw on the output, H remains accountable for ensuring correct medical outcomes. Thus, H needs transparency regarding the functionality of the AI system.

By relying on global model-agnostic methods, H gains a broad understanding of the AI system and its overall behavior. Together with the help of E, global model-agnostic methods help H to assess whether the recommendations of the AI system align with their expectations and medical knowledge. However, while global model-agnostic methods help H (and E) to gain an overall sense of the AI system, H (and E) are usually confronted to explain and justify individual decisions. Thus, global model-agnostic methods are usually too broad to explain and justify the behavior of the AI system in specific situations. Consequently, D must also equip H (and E) with local model-agnostic methods, like LIME (Ribeiro et al. 2016).

Local model-agnostic methods, like LIME, offer H (and E) a detailed explanation of specific recommendations for the AI system (Molnar 2019). By analyzing individual recommendations in detail, H can determine whether the recommendations of the AI system are logical and consistent with known medical causal relationships. This fine-grained transparency empowers H to incorporate its medical expertise and make informed decisions about diagnosis and medical treatment, respectively to explain and justify its decisions when false diagnoses or medical treatments occur. This helps H to fulfill its accountability obligations, even in cases of erroneous outputs.

5. Outlook

We acknowledge that our proposed model looks complicated. There are at least four to five parties involved: patient, healthcare professional (plus perhaps a clinical AI expert), developer and oversight authorities – and, of course, the AI system in between. While the distinctions between those parties are based on functional roles, it is theoretically possible for a healthcare professional to act as their own clinical AI expert or developer, though this is unlikely in practice. We do not claim that the division of accountability does work, but we claim that the regulation of the relationships between those parties is required if it is still possible to address questions of accountability appropriately to avoid an accountability diffusion. In the case of cancer diagnoses, the disappearance of accountability implies a loss of patient rights. A patient would not be able to address his or her claim for appropriate treatment to an instance that is expected to fulfill this claim and which is able to do so.

In that context, it is important to emphasize that the model is not meant to be static. We just distinguish the different instances and their role obligations. But that does not mean that the relationship is a static one. It is also possible that healthcare professionals learn to use AI systems as part of a practice where they are still in charge, but it is likewise possible that they become increasingly dependent on the systems and lose their own diagnostic capacities and the ability to rely on their own judgment. It is not decided in advance how the future will develop. However, the central point is that the advantages of AI systems can only be utilized at the cost of more involved parties and with more complex relationships that have to be evaluated. However, we propose that using XAI methods can help each party fulfill the accountability obligations that arise at their specific touchpoints with the AI system. In this vein, our model allows us to assess whether the advantages of the use of AI systems are worth the introduction of more complex accountability relationships and how XAI methods can help in each phase. But in any case, if it is responsible to introduce AI systems, then it must be seen as a necessary requirement that there is still a functioning system to hold human actors accountable vis-à-vis the patient to ensure patient rights. This implies that the accountability of the healthcare professional would still be the cornerstone of medical ethics and medical law. Additionally, it would still be possible for one healthcare professional to speak with one patient from person to person, listen to the wishes and fears, the hopes and expectations of this person, and on the basis of this conversation, propose to this patient the best treatment for this person. The introduction of AI systems should not undermine this conversation from one person to another. It should still be possible that the patient accuses the healthcare professional of giving wrong advice. All of this is required so that medical law and medical ethics are still functioning.

References

Adadi, A. and Berrada, M. (2018): “Peeking Inside the Black-box. A Survey on Explainable Artificial Intelligence (XAI)”, in: IEEE access, 6, pp. 52138–52160.

Apley, D. W. and Zhu, J. (2020): “Visualizing the Effects of Predictor Variables in Black Box Supervised Learning Models”, in: Journal of the Royal Statistical Society Series B: Statistical Methodology, 82, pp. 1059–1086.

Asatiani, A. et al. (2020): “Challenges of Explaining the Behavior of Black-box AI Systems”, in: MIS Quarterly Executive, 19, pp. 259–278.

Beeravolu, A. R. et al. (2021): “Preprocessing of Breast Cancer Images to Create Datasets for Deep-CNN”, in: IEEE Access, 9, pp. 33438–33463.

Berente, N. et al. (2021): “Managing Artificial Intelligence”, in: MIS Quarterly, 45, pp. 1433–1450.

de Boer, B. and Kudina, O. (2021): “What is Morally at Stake When Using Algorithms to Make Medical Diagnoses? Expending the Discussion Beyond Risks and Harms”, in: Theoretical Medicine and Bioethics, 42, pp. 245--266.

Brancaccio, G. et al. (2024): "Artificial Intelligence in Skin Cancer Diagnosis. A Reality Check", in: Journal of Investigative Dermatology, 144, pp. 492–499.

Breiman, L. (2001): "Random forests", in: Machine learning, 45, pp. 5–32.

Bringsjord, S. (2008): "Ethical Robots. The Future Can Heed Us", in: AI & Society, 22, pp. 539–550.

Cooper, A. F. et al. (2022): "Accountability in an Algorithmic Society. Relationality, Responsibility, and Robustness in Machine Learning", 2022 ACM Conference on Fairness, Accountability, and Transparency, Seoul, Republic of Korea.

Cunningham, P., Cord, M. and Delany, S. J. (2008): Supervised learning. Machine Learning Techniques for Multimedia. Case Studies on Organization and Retrieval, Springer.

Düwell, M. (2012): Bioethics. Methods, Theories, Domains, Routledge.

Dwivedi, R. et al. (2023): "Explainable AI (XAI). Core Ideas, Techniques, and Solutions", in: ACM Computing Surveys, 55, pp. 1–33.

Fisher, A., Rudin, C. and Dominici, F. (2019): "All Models are Wrong, but Many are Useful. Learning a Variable's Importance by Studying an Entire Class of Prediction Models Simultaneously" in: Journal of Machine Learning Research, 20, pp. 1–81.

Fiske, A., Henningsen, P. and Buyx, A. (2019): "Your Robot Therapist Will See You Now. Ethical Implications of Embodied Artificial Intelligence in Psychiatry, Psychology, and Psychotherapy", in: Journal of Medical Internet Research, 21, e13216.

Friedman, J. H. (2001): "Greedy Function Approximation. A Gradient Boosting Machine", in: Annals of Statistics, 29, pp. 1189–1232.

Gilvary, C. et al. (2019). "The Missing Pieces of Artificial Intelligence in Medicine", in: Trends in Pharmacological Sciences, 40, pp. 555–564.

Goldstein, A. et al. (2015): "Peeking Inside the Black Box. Visualizing Statistical Learning With Plots of Individual Conditional Expectation", in: Journal of Computational and Graphical Statistics, 24, pp. 44–65.

Gupta, M. and Gupta, B. (2018): "A Comparative Study of Breast Cancer Diagnosis Using Supervised Machine Learning Techniques", Second International Conference on Computing Methodologies and Communication (ICCMC), IEEE, pp. 997–1002.

Hariharan, S. et al. (2023): "XAI for Intrusion Detection System. Comparing Explanations Based on Global and Local Scope", in: Journal of Computer Virology and Hacking Techniques, 19, pp. 217–239.

Hunter, D. et al. (2012): "Selection of Proper Neural Network Sizes and Architectures – A Comparative Study", in: IEEE Transactions on Industrial Informatics, 8, pp. 228–240.

Jiang, F. et al. (2017): "Artificial Intelligence in Healthcare. Past, Present and Future", in: Stroke and Vascular Neurology, 2, pp. 230–243.

Krogh, A. (2008): "What are Artificial Neural Networks?", in: Nature Biotechnology, 26, pp. 195–197.

Kudina, O. and de Boer, B. (2021): "Co-designing Diagnosis. Towards a Responsible Integration of Machine Learning Decision-support Systems in Medical Diagnostics", in: Journal of Evaluation in Clinical Practice, 27, pp. 529–536.

Lee, S. S. (2022): "Philosophical Evaluation of the Conceptualisation of Trust in the NHS' Code of Conduct for Artificial Intelligence-driven Technology", in: Journal of Medical Ethics, 48, pp. 272–277.

Linardatos, P., Papastefanopoulos, V. and Kotsiantis, S. (2020): "Explainable Ai. A Review of Machine Learning Interpretability Methods", in: Entropy, 23, 18.

Matthias, A. (2004): "The Responsibility Gap. Ascribing Responsibility for the Actions of Learning Automata", in: Ethics and Information Technology, 6, pp. 175–183.

Molnar, C. (2019): Interpretable Machine Learning.A Guide for Making Black Box Models Explainable, Lulu.com, self-published.

Morley, J. et al. (2020): "The Ethics of AI in Health Care. A Mapping Review", in: Social Science & Medicine, 260, 113172.

Nassif, A. B. et al. (2022): "Breast Cancer Detection Using Artificial Intelligence Techniques. A systematic Literature Review", Artificial intelligence in medicine, 127, 102276.

Osareh, A. and Shadgar, B. "Machine Learning Techniques to Diagnose Breast Cancer", 2010 5th international symposium on health informatics and bioinformatics, 2010. IEEE, pp. 114–120.

Perincheri, S. et al. (2021). "An Independent Assessment of an Artificial Intelligence System for Prostate Cancer Detection Shows Strong Diagnostic Accuracy", in: Modern Pathology, 34, pp. 1588–1595.

Ponnada, V. T. and Srinivasu, S. N. (2019): "Edge AI System for Pneumonia and Lung Cancer Detection", in: International Journal of Innovative Technology and Exploring Engineering, 8, pp. 1908–1915.

Retzlaff, C. O. et al. (2024): "Post-hoc vs Ante-hoc Explanations. xAI Design Guidelines for Data Scientists", in: Cognitive Systems Research, 86, 101243.

Ribeiro, M. T., Singh, S. and Guestrin, C. (2016): "'Why Should I Trust You?' Explaining the Predictions of Any Classifier", Proceedings of the 22nd ACM SIGKDD International Conference on Knowledge Discovery and Data Mining, pp. 1135–1144.

Rudin, C. (2019): "Stop Explaining Black Box Machine Learning Models for High Stakes Decisions and Use Interpretable Models Instead", in: Nature Machine Intelligence, 1, pp. 206–215.

Rudschies, C. and Schneider, I. (2024): "Ethical, Legal, and Social Implications (ELSI) of Virtual Agents and Virtual Reality in Healthcare", in: Social Science & Medicine, 340, 116483.

Russell, S. J. and Norvig, P. (2016): Artificial intelligence: a modern approach, Pearson.

Schleidgen, S. and Friedrich, O. (2023): "Künstliche Intelligenz in Medizin und Pflege", in: Ethik in der Medizin, 35, pp. 169–172.

Sechopoulos, I., Teuwen, J. and Mann, R. (2021)"Artificial Intelligence for Breast Cancer Detection in Mammography and Digital Breast Tomosynthesis. State of the Art", in: Seminars in Cancer Biology, pp. 214–225.

Shin, D. (2021): "The Effects of Explainability and Causability on Perception, Trust, and Acceptance. Implications for Explainable AI", in: International Journal of Human-Computer Studies, 146, 102551.

Shravya, C., Pravalika, K. and Subhani, S. (2019): "Prediction of Breast Cancer Using Supervised Machine Learning Techniques", in: International Journal of Innovative Technology and Exploring Engineering (IJITEE), 8, pp. 1106–1110.

Speck, H. et al. (2021): Digitalisierung im Gesundheitswesen: anthropologische und ethische Herausforderungen der Mensch-Maschine-Interaktion, Herder.

Suzuki, H. et al. (2021): "Artificial Intelligence for Cancer Detection of the Upper Gastrointestinal Tract", in: Digestive Endoscopy, 33, pp. 254–262.

Tosun, A. B. et al. (2020): "Explainable AI (xAI) for Anatomic Pathology", in: Advances in Anatomic Pathology, 27, pp. 241–250.

Zheng, D., He, X. and Jing, J. (2023): "Overview of Artificial Intelligence in Breast Cancer Medical Imaging", in: Journal of Clinical Medicine, 12, 419.

Trust in AI: A Unified Approach

Andreas Kaminski

Abstract *The discourse on trust in technology, and especially in AI, has crystallized around two dominant positions. According to the first, technology or AI cannot be trusted and doing so would constitute a category mistake. In response, a second position has emerged, which interprets trust in technology as reliability. This represents, in effect, a retreat: it responds to the critique of the first position by adopting a significantly narrower epistemic concept of trust. This paper acknowledges the initial critique (i.e., the category mistake) as valid. However, it argues that retreating to a reliabilist understanding of trust is not the only available option. An alternative remains open, one that allows for a richer concept of trust, grounded in virtue theory. This alternative shifts the focus to the relationship between individuals, organizations, and technology. The paper outlines the basic contours of this alternative approach to understanding trust in AI.*

Trustworthiness of AI as well as trust in AI has become a title of numerous publications (AI HLEG 2019; Floridi 2019; Thiebes, Lins and Sunyaev 2021). The title presupposes that talking about trust in technology makes sense which is the subject of ongoing debates (Wagner 1994; Nickel, Franssen and Kroes 2010; Kaminski 2010; Floridi 2019; Thiebes, Lins and Sunyaev 2021). This has been doubted, and for good reasons (cf. Lahno 2002: 134–138). Some theories of trust have presented cogent arguments that this way of speaking constitutes a category mistake; that is: applying the concept of trust to technology in general and AI in particular is fundamentally misguided. According to these arguments, speaking of trust in technology is as nonsensical as attributing diabetes to a number or declaring a rock on the moon bankrupt. At best, we could try to read such statements metaphorically.

Given these arguments, it seems that any defense of trust in technology must rely on a substantially deflated or impoverished version of the concept of trust. A natural course, then, appears to be limiting the concept to a specific epistemic variant. Reliabilist approaches in particular appear to offer a viable path for reconciling the critique with a coherent way of talking about trust (Durán and Formanek 2018). On this understanding, a system is reliable when, in most cases, it produces the desired effect or serves the intended purpose. Reliabilist approaches therefore construe the trustworthiness of systems primarily as something that can be recognized and even measured – above all by their track record. This comes with the aforementioned limitation: Trust is understood in a different, restricted sense within these epistemic theories. The limitations include that the specifically normative dimension of trust and the special form of relationship that exists between persons who trust each other cannot be captured. When trust is disappointed, the response is mainly (or at least also) morally determined. The epistemic approach, however, cannot make sense of this. On that view, whether we trust or not depends solely on our epistemic expectations.

In trust research, it has become something of a standard move to treat trust as a multifaceted social phenomenon, requiring different theoretical accounts depending on the context – whether between friends, market actors, or in relation to God or technology. From this standpoint, it seems reasonable to propose a distinct concept of trust that, while not capturing all phenomena, at least makes sense of talking about technology.[1]

In the following, I want to show three things:

1) The arguments that a certain way of speaking about trust in technology constitutes a category mistake are compelling;
2) Nevertheless, there is a different understanding of what it means to say that we trust technology – one that does not amount to a category mistake;
3) Moreover, this way of speaking avoids a reductive understanding of trust but maintains its normative significance.

I begin by outlining the basic features of epistemic trust theories. Subsequently, I address the criticism of epistemic trust theories, which simultaneously forms the basis for the arguments why trust in technology is not possible. Then I elaborate on the fallback position that trust in technology should therefore be understood merely epistemically. The following section will then present the alternative position. Finally, I outline what this means for responsibility and the capacity-based approach.

1 This routine was able to spread easily within the humanities and social sciences since theoretical pluralism is taken for granted in these disciplines. Moreover, there are specific reasons tied to the theoretical history of trust. At the beginning of the current wave of intense engagement with trust, Annette Baier raised a criticism in an essay that, in many ways, prefigured later debates: Namely, that when trust had been a subject of philosophical reflection at all, it had typically been understood through the model of a contract (Baier 1986). That is, as a situation conceived symmetrically, where two equal parties enter into an explicit agreement. This implicit orientation toward the contract model, however, is ill-suited to many other forms of trust – for instance, those between parents and children. A plurality of forms of trust, it was argued, should be matched by a plurality of theories of trust (cf. Frevert 2013: 9; Hardin 2002: 6; Hartmann 2001: 7 et seq.).

Epistemic Theories of Trust

Epistemic theories of trust come in various versions: the so-called evidential view (examples are Hume 2007; Coady 1992) or rational choice and game theory (Coleman 1990; Gambetta 1988; Hardin 2006). What they have in common is that they conceptualize trust as a relationship between A and B in which A *knows* something about B that leads A to expect that B will act in a way that benefits A.

In the empirical version of the evidential view, for instance, for David Hume (Hume 2007, ch. X) or in the psychology of testimonial assertions (Stern 1903; Sporer 1997; Arntzen 2007), this knowledge on A's part might consist of B's track record of acting in a trustworthy manner. It could also be based on indicators that show whether B will act in a trustworthy way. A prime example would be credit models (Lauer 2017). These provide A with a track record, derived from experience, showing how often B has taken out loans and repaid them. A can also use indicators to classify B into a (statistical) group for which there is a known track record; for instance, B's place of residence, level of education, age, number of children, or the presence of a car (kept in a garage). According to this approach, entirely different indicators could also be used, such as those analyzing B's facial expressions as signs of truthfulness or a deceptive intent in communication (Ekman 2001, 2003; Kaminski 2020b).

According to this theory, trust is based on a foundation that is *independent* of trust itself. This foundation is the evidence for B's trustworthiness or lack thereof: A *knows* whether B is trustworthy. Trust is then the inference drawn from this evidential basis: a forecast giving rise to a cognitive expectation. When conceptualized in this way, trust becomes indistinguishable from mere prediction (of a positive outcome). Such a concept of trust or a method of establishing trustworthiness can be applied to persons, institutions, plants, animals, technology, etc. The underlying reason is that, essentially, we are dealing only with *inductive inferences*[2] that are understood from a *practical perspective*, insofar as they guide decisions. Trust is the natural conclusion when the expectation is that a positive outcome will occur; conversely, if a positive outcome is not anticipated, the result is distrust. From this, we can already provide a characterization of what trust means in this epistemic version: *Relying on the other because one relies on the procedure.* Primarily, it is *not* the other person (or the technology depending on what occupies the other position in the relationship) that is trusted, but the evidence itself and the process of drawing correct conclusions from it. One relies on B by, for example, relying on the reliability of the indicators. The counterpart's reliability depends on the reliability of the (inductive) procedure. Naturally, different indicators can have varying degrees of reliability.

---

2 This was already clear to Hume. See his remark in the *Enquiry*, where he notes that these inferences are the same as those drawn when discovering the connection between cause and effect (Hume 2007: 80).

Cognitive expectation, cognitive disappointment: Because it is strictly a cognitive expectation, disappointment is similarly cognitive in nature. One might just be surprised or respond with understanding if the expectation was never very certain in the first place, since disappointment can be expected in such cases (at roulette, sometimes the ball lands on red, sometimes on black). Within this framework, one can never be *morally* disappointed.[3]

More evidence, more trust: If a positive outcome can be expected, more evidence leads to greater trust. It does not matter what the evidence is based on or how it is obtained: Threats of sanctions or surveillance could, in principle, boost trust just as much as the other person's virtues.

Other variations of epistemic theories, such as rational choice or game theory, remain within the same framework. The main characteristics are the same. Additional assumptions may appear, for example (i) regarding the other person's interests, (ii) that the other person will pursue these interests rationally, and (iii) in game theory in particular, calculations that compare one's own expectations with those of others. Nevertheless, these additions do not alter the epistemic framework; they simply refine it. Thus, Coleman can regard the question of whether one should trust others as a special case of decision-making under risk, conceptualized in the form of a *bet* (Coleman 1990: 99). For rational choice and game theorists, the trustworthiness of the other person ultimately reduces to a probability value (Dasgupta 1988; Gambetta 1988).

---

3 The reason for this is not that one is dealing, for instance, with economic relations, but rather that the theory itself lacks a framework for articulating a moral violation. If an evidence-based prediction is made and the claim contained in the prediction does not come true, the person who made the prediction cannot be morally disappointed that the object whose behavior was predicted did not act as expected. The reference to the object (an evidence-based prediction) does not allow for the formulation of a moral disappointment. This holds regardless of whether the prediction concerns the course of the stars or the behavior of people. Only when the reference is no longer purely epistemic can the possibility arise for expressing moral disappointment. Now, consider a case in which a person enters into an economic cooperation based on the game-theoretic prediction that their potential partner will act cooperatively – and is then disappointed. In this case, game theory does not provide grounds for moral disappointment. At most, the disappointment could serve as a kind of sanction, which in turn constitutes an epistemic reason; namely, a factor intended to promote current or future cooperative behavior. This can be taken into account in future predictions, in the sense that the threat of sanction is treated as a motivating factor that increases the likelihood that the other person will act cooperatively. To be sure, people who reason in game-theoretic terms can in fact experience moral disappointment. But they can not do so on the basis of game theory. Their moral disappointment rather reveals that they relate to the other person in a way that is not purely epistemic.

Critique of Epistemic Theories of Trust

Within epistemic theories, the role of evidence and the resulting forecasts leads to a conception of trust that does not fundamentally differ from predictions about the behavior of stars. One 'trusts' that a certain star will appear in the same place tomorrow, just as one trusts that a friend will show up in time to pick one up to go to the airport. Ultimately, as we have seen, trust is not placed in the other person but in one's own cognitive faculties, which have carefully evaluated the evidence. This basic situation does not change even if we factor in second-order expectations as in game theory. Such considerations merely complicate inference but do not alter the nature of the relationship between A and B.

The critique of epistemic trust theory, advanced, for instance, by the so-called *assurance view* or by theories that treat trust as a feeling, therefore focuses on the nature of the relationship. I highlight three criticisms, which form a family of objections insofar as they all center on this relational aspect:

1. *Not actually trusting the other person:* In epistemic theory, trust in a car or in a friend can be assessed in essentially the same way: Both could, for example, have a good track record. The car always starts, and the friend always shows up when needed. The reason is that, within epistemic theory, the other person ultimately is not considered *as a person*. Indeed, the more thoroughly one understands the rules that govern another's actions, the better one can assess their trustworthiness. These rules gain in evidential force the more they resemble identifiable mechanisms (Lahno 2002: 168–170). However, once another's actions are interpreted merely as the outputs of an underlying mechanism, the person is no longer seen as an autonomous agent but rather as analogous to a natural process or technical system. In contrast, non-epistemic accounts emphasize the autonomy of the other as the foundation of their trustworthiness (Lahno 2002: 169–178; Moran 2006: 278 et seq.).

2. *The normativity of trusting relationships:* If trust ultimately rests on the reliability of a mechanism (producing a merely cognitive expectation), then its distinct normativity cannot be adequately captured. A person whose trust has been betrayed is not simply surprised in a factual sense but wronged in a normative one. The disappointment is not just interpreted as the truster's mistake, e.g., an overestimation of the available evidence – that is possible too – but at least also as a failure on the part of the trusted person. What is at stake is not merely the expectation that p will occur, but the expectation that you will ensure that p. I do not merely expect *that* p, I expect *from you* that p (Faulkner 2007: 881). In this light, disappointment becomes a breach of trust, a "betrayal" (Lagerspetz 1998: 42). Trust presupposes that the other person assumes responsibility voluntarily and autonomously within the relationship (Moran 2006: 281). Only what someone takes on intentionally and freely can genuinely be the object of trust.

3. Impermissible epistemic reasons: Because epistemic theories ground trust in epistemic reasons, certain consequences arise that run counter to the very nature of trust. Increasing sanctions or threats of sanctions can strengthen these epistemic reasons, which, within the epistemic perspective, ought to enhance trust. Likewise, from an epistemic point of view, surveillance would justify "more trust", insofar as it provides better informational grounds.[4] The same essentially applies to increasing control over the other person. Someone who reads another person's diary, monitors them at every turn, and threatens them with severe penalties will, from an epistemic standpoint, have a better basis for their 'trust' (Kaminski 2020a: 181; 2023). Clearly, if trust is supposed to rest on a voluntary assumption of responsibility, such epistemic grounds are excluded.

The fundamental error in all three cases can be traced back to the same category mistake: Epistemic theories use categories suited to causal prediction of natural or technical processes to describe interpersonal relations grounded in autonomy. They base their understanding of trust on procedures that do not treat others as autonomous persons but rather assess or measure their 'trustworthiness' in light of their predictability analogous to mechanisms. Trust, by contrast, requires understanding others as autonomous persons. If the freedom of the other is not placed at the center, predictions – especially those grounded in technomorphic forms of control – are easily mistaken for trust. This does not mean that others cannot be reliable. One can form expectations, but these rest on fundamentally different grounds: not on mechanical patterns of behavior, but on the other person's autonomous decision to be trustworthy.

The 'Reliability Residuum' and Its Problems

In light of the aforementioned critique, it may seem that there is no longer a meaningful way to speak of trust in technology and thus in AI systems. The criticism of the epistemic theory of trust showed that the form of the relationship between two individuals who trust each other cannot be replicated by technical instruments, machines, or systems. For such replication to occur, technology would have to be an autonomous subject capable of recognizing and voluntarily entering into normative

---

4 More precisely put: This applies in cases where surveillance reveals that the other person intends to cooperate, harbors no harmful intentions, and so on. However, the central point remains unaffected: Trust and direct knowledge are mutually exclusive. If I watch the other person and *see* that they are attending to my concerns, I no longer need to trust them. The same holds for control: Trust and control are mutually exclusive. If I control the other person's behavior – or even merely attempt to do so (for instance, by threatening sanctions) – then I am not trusting them. From within the epistemic approaches, however, there is no clear reason why such surveillance or control would not serve to justify (or even promote) trust.

commitments. Someone who trusts a given technology would thus not merely base their trust on knowledge about that technology; they would base it on the relationship they have with the technology. In that relationship, they would not only expect a certain effect to occur when initiating a process, but also expect the technology to care that this effect occurs – precisely because it is trusted.

It is obvious that technology does not and cannot fulfill this condition. Otherwise, one would have to assume that technology enjoys the status of an autonomous subject that does not merely operate according to rules we can discern but also can place itself in a normative relationship by acknowledging normative rules and acting in accordance with them. Situations in which technology is sometimes assumed to meet this condition appear to arise from an ambiguity – namely, equating the moral autonomy of persons with the technical autonomy of systems, whose functioning may not be determined directly by developers but is governed by higher-level rule systems that generate the operative rules only afterward (Kaminski 2014).[5] However, this question cannot be pursued further here, so I will proceed on the assumption that technical systems do not exhibit moral autonomy.

If this is the case, then only a weaker sense remains in which one could speak of 'trust' in technology. And that sense seems to be delineated precisely by the epistemic theory. Since the latter does not capture trust in a substantive sense, a restricted concept must take its place. Hence, the discourse has turned to the reliability of technical systems. On the one hand, reliability was introduced, albeit in an unclear manner, by Annette Baier as the opposite of trust in a narrower sense at the start of the modern debate (Baier 1986). On the other hand, reliability is also a familiar technical concept. For instance, AI models have long been tested to see how reliably they carry out specific classifications like identifying objects in images; currently large language models are assessed in terms of their reliability in generating accurate responses.

Here we rediscover Hume's idea of a track record: The degree of trustworthiness we ascribe to someone who promises something is evaluated by the ratio of total promises made to promises actually fulfilled. Analogously, the degree of reliability of image recognition is determined by comparing the total number of image-recognition processes with the number of correct recognitions.[6]

Should we therefore confine our talk about the trustworthiness of technology to this notion of reliability? Must we do so, given the arguments that rule out any

---

5 In the case of machine learning algorithms, the base architecture, hyperparameters (such as the learning rate), cost function, and data are specified by developers but not the learned model itself; that is the result of the training process. As a consequence, what emerges is a kind of program that was not directly coded (Angius and Plebe 2023).

6 In AI literature, the terminology may differ insofar as there is also talk of "robust accuracy". However, this is merely a terminological difference, not a conceptual one.

broader understanding? One initial problem is that many debates on the *trustworthiness* of AI cannot be limited to whether the technology reliably detects and classifies objects in data. While this is *one* important dimension, it is by no means the only relevant factor for deciding whether a model is trustworthy. An AI system that consistently reaches unjust decisions would hardly be considered trustworthy; likewise, a chatbot that reliably deceives users would not merit that label – at least not by those who consider justice or veracity important. A system that is indeed reliable and just but fails to protect sensitive data would likewise not be deemed trustworthy.

Hence, we encounter two claims that appear to clash:

1. We can talk about technology as reliable, but there is no meaningful sense in which we can call it trustworthy.

2. Reliability is not sufficient for how we assess the trustworthiness of technology in our discussions.

One could attempt to resolve this terminologically. But that would be unlikely to succeed, for, as point (2) shows, there is a significant way of talking about technology that does not limit itself to reliability. Faced with this conceptual dilemma, we initially have two options:

a. We can either try to reduce all aspects of trustworthiness (2) to reliability (1).
b. Or we can take the opposite approach and differentiate the layers, attributing (1) to the technical sphere and (2) to the social sphere.

However, neither option seems promising.

Let us start with the idea of *differentiating* levels: The solution would be to say that we attribute reliability to the technical system while ascribing broader trustworthiness – such as meeting standards of justice or protecting sensitive data – within the social dimension; presumably, this would mean that we trust the developers or providers of these systems. This may not be entirely wrong, but it oversimplifies the fact that we do, in fact, trust the system itself to make fair decisions or adequately protect data. After all, it is not the developers themselves who make these decisions in real time; the system must provide the security measures that prevent unauthorized access to the data it contains.

The alternative solution would be a *reduction*: Justice or data privacy (or privacy more generally) would be regarded as special cases of system reliability. One could flesh out this idea as follows: A system is fair if it reliably produces fair decisions; it protects data if it reliably ensures that only authorized data accesses occur. In this way, reliability would become the overarching framework for evaluating systems (and people) in numerous dimensions. What we previously labeled reliability – the system's functional correctness, for instance – would itself be just one special case (which one might call functional reliability or robust accuracy). If the function of the system is to classify objects in image data, one can then say the system is reliable to such-and-such a degree at fulfilling that function.

A central difficulty with this approach, however, is that interpreting justice or data protection as mere reliability can gloss over serious violations by focusing only on the quantity of failures rather than their quality. A system that is "nearly always" fair or that "nearly always" protects data but occasionally allows severe transgressions would, by this definition, still be a reliable system – even if the exceptions were catastrophic. One might try to compensate for this by taking the severity of individual cases into account, but then one would be leaving behind the tightly circumscribed lines of a reliabilist approach, since it would require engaging in substantive debates about, for example, the nature of justice. Such an attempt at stabilization would naturally lead to what I am about to present.

Practices and Outcomes in the Light of Virtues

How can we resolve the problem of trust in technology, given the dead ends reached by the earlier proposals? We have seen that talk of trust in a substantive sense goes beyond mere reliability and that it seemingly applies only to autonomous persons – those who understand that they are trusted, who voluntarily enter into trusting relationships and who can acknowledge the normative commitments and expectations these relationships entail. Yet the same observation appears to hold true for the predicate 'just'. Much like the term 'trustworthy' seems to presuppose personhood, one might assume that 'just' is likewise meaningful only when applied to persons. But this is not in fact the case: We speak of just rules without thereby granting the rules themselves a subject status in this sense. Should we, by the same logic, restrict our talk of 'just' in the same manner suggested by the above argument concerning 'trustworthy'?

My thesis is that precisely the opposite is true. We can learn something about the trustworthiness of technical systems from the way we say that certain rule systems are just or unjust. When we call a system of rules just, we mean more than that it reliably produces just decisions (although that is a necessary condition, as established in Section 3, it is not sufficient). Nor do we ascribe subject status to them. Rather, I suggest that what we mean is that these rules result from deliberations conducted in the spirit of justice – deliberation about, say, the fair distribution of resources. The rules were developed with the virtue of justice in mind. They are not accidentally just; rather, they were deliberately shaped to embody a 'just spirit'.

Another way to understand this proposal is to see it as drawing on the idea of a *pros hen*-structure.[7] In other words, there is a focal meaning of discourse about virtues: Virtues primarily concern a person's character. Yet we can also say of a person's actions that they are just or unjust and of the consequences of those actions that they result in just or unjust outcomes. By the same token, we can speak of AI systems and technology in general as manifesting certain virtues.

This proposal will (and should) prompt a series of follow-up questions. These concern the relationship between persons and technical systems, the connection between trustworthiness and other virtues, the epistemic and normative dimensions of trustworthiness, and finally, how one might establish whether something is in fact virtuous.

Let us begin with the question of how persons and technical systems relate. In most cases, we do not know the developers of technical systems personally. How, then, can we say anything about the systems' virtue as derived from their developers? Put differently, if we describe a technical system as trustworthy while grounding that concept (via a *pros hen*-structure) in the notion of a person's character, yet lack any knowledge of the persons who developed or deployed the technical systems, how meaningful can it be to speak of 'trustworthy systems' at all? Here again, thinking about just rules offers a helpful analogy. To judge the justice of rules, we need not know the people who developed them. We can analyze the rules themselves and conclude whether they are just or unjust. An example would be gerrymandered voting districts – drawn so that one group's votes count disproportionately. Determining whether voting districts are unjustly structured does not require knowledge of who drew them. Such knowledge can, however, serve two purposes: First, it might prompt us to investigate further, if we know these particular individuals have engaged in such practices before (or used other unfair campaign tactics). Second, it may be needed to understand whether the injustice was a lapse or mistake, the result of incompetence, or a malicious, intentional attempt to make certain votes 'count more'. The same holds for AI systems. Whether a system produces fair decisions ultimately must be shown by examining those decisions themselves. Deciding whether an unfair system was simply sloppily developed or was intentionally designed to be that way may indeed involve knowing more about the developers – though in some cases, the system's very structure might reveal enough on its own.

7 Aristotle develops this idea about the structure of central concepts in our language in his *Metaphysics* with reference to the manifold ways in which we speak of being (cf. Aristoteles 1994: IV.2, 1003a2, 1003a33–1003b19). *Pros hen* could be translated as "in relation to one". For instance, there are various meanings of the term 'healthy' or 'health'. We speak of healthy food and of a person being healthy. In the former case, "healthy" refers to a cause: food that promotes health while in the latter, it denotes a result or state. According to Aristotle, these meanings are not unrelated. Rather, they can be organized with reference to a focal meaning: the healthy body. It is this central case from which the other uses derive their meaning.

The core idea, then, is this: We can trust technical systems when we view them as having been developed in the spirit of trustworthiness. That is, they were created with the explicit aim of making them trustworthy. In adopting this perspective, we are shifting our frame of reference: In evaluating trustworthiness, what we truly assess is whether the system's development was guided by the values that should be central to creating trustworthy technologies. The systems are the result of practices directed by developers who wanted to produce trustworthy systems. The analogy to rules that emerged from processes oriented toward the virtue of justice is illuminating here as well. However, acknowledging the significance of development processes (and thus the developers themselves) does not imply that people can trust only those systems whose developers they personally know. The very design of the systems can bear the marks of orientation toward such virtues, and this is what our trust rests on.

Imagine someone who is analyzing a technical system and finds that it was designed in such a way as to make it particularly difficult to hack. In such a case, it is of course appropriate to say that the system was designed with a focused concern for the value of security – that it was developed in that spirit. Someone who speaks this way need not be presumed to have any personal acquaintance with the system's developers.

This raises the question of who is in a position to analyze a system in such a way as to determine which values it embodies and which values, therefore, informed its design. When I look at a chair, I might still have a few cues that help me judge whether it was designed to be robust, comfortable, or whether a low price point was the primary consideration. Many other aspects, such as ecological values, tend to elude direct assessment. This is all the more true for information technology, and especially for complex digital systems: Our capacity for evaluative judgment becomes markedly more constrained when we turn to AI models. To evaluate the values according to which such systems are designed, we are far more dependent on the expertise of others. One can nonetheless see the general direction that an assessment of trustworthiness might take. However, within the context of this article, the question is not how such an assessment could be carried out in practice. Rather, the focus is on the theoretical question: Is there a meaningful way to speak of trust in technology in general, and trust in AI in particular? I believe I have now sketched the essential contours of the answer. The task now is to develop this outline in more detail.

Trustworthiness as the Unity of Other Virtues

What kind of virtue is trustworthiness? When we consider trustworthiness alongside other values such as justice, privacy, and reliability, each value appears to address a specific aspect of the system: Reliability measures how consistently a system performs its intended function, privacy concerns how well data is protected against unauthorized access; justice assesses whether decisions, for example, about the allocation of resources, are made in a way that ensures no individual is disadvantaged.[8] Is there a particular aspect devoted to the trustworthiness of systems, alongside other aspects such as justice or privacy? Initially, it might look as though this aspect is empty, as though it cannot be filled with any concrete content. But this perception stems from the assumption that trustworthiness is just one additional segment – expressed in requirements-engineering terms, one non-functional requirement among others.

The confusion disappears if we look at how breaches of trust are typically discussed: "M was not honest", "N did not have the courage to stand up for me", "O did not care that their decision was unjust". Such statements might explain why someone lost trust in another. If we consider these explanations closely, we see that trust relates to other values. For instance, when we trust someone, we often trust that a friend will be *honest*, that a colleague will be *sincere*, that a partner will *courageously* stand up for us, that a doctor will be *careful*, that a supervisor will be *fair*, and so on. From this follows the idea that trust does not stand alongside other values or virtues but is instead linked with them in the following way: *The object of trust is the other person's trustworthiness, and trustworthiness in turn relates to other values or virtues.* A trustworthy person will prove to be honest, just, courageous, and kind when it matters. In short, trustworthiness is the unity of other virtues; it is not just one virtue among many, but rather the social connection point at which they come together (Kaminski 2020a: 396 et seq.)[9]

By analogy, the trustworthiness of technical systems is not merely one non-functional requirement among others such as privacy; rather, it is their overall unity. When we trust technical systems, we trust them to be trustworthy in all the relevant areas; the relevant areas are precisely those values we consider essential – such as ensuring no discriminatory bias, avoiding unjust recommendations, protecting data, reliably performing their function, and so forth. From this perspective, it is hardly surprising that lists describing "trustworthy AI" often lack a clear systematization of the values at stake. Trustworthiness is oriented toward the specific values and virtues that matter in each particular context. Which ones matter depends on

8 See, for example, the lists of values such as the requirement that technical systems be "effective, interoperable, robust, and reliable" (Art. 50, No. 2, AI Act), or that they ensure "accuracy, reliability, and transparency" in order "to avoid adverse impacts, retain public trust" (Recital 59, AI Act).

9 This raises the question of how this thesis relates to Aristotle's idea of justice as a unity. If we understand justice as the universal appropriateness of one's actions, then this apparent conflict may be smaller than it now seems (cf. Aristoteles 2011: 1129a).

the situation, something that, as Aristotle remarks, can be specified only in general outlines (Aristoteles 2011: 1104a).

Even though the formal object of trust is thus the trustworthiness of persons or technical systems, and that trustworthiness points toward other values or virtues, this does not automatically mean that trust must always be understood as a three-place relation: A trusts B to do p. Granted, a form of trust that never manifests in any concrete relation (p) would not only be unrecognizable but would eventually cease to exist. I see the relationship between the three-place and two-place formulations of trust as a modal relation (actual vs. possible). Trust does manifest itself concretely in particular instances, but it is not reducible to them since the trust relationship creates (and also delimits) the possibility of various future scenarios. And precisely because situations can only be anticipated in broad outlines, it is impossible to specify exhaustively in a finite list all the things our trust might someday encompass.

The Epistemic and Normative Dimension of Trust

The notion of trust in technology introduced here, which, from my perspective, no longer entails a category mistake, goes hand in hand with a different theory of trust. Besides the epistemic theory of trust (in its various forms), there is a well-established normative view of trust. This normative view appears in different guises: the assurance view (Moran 2005, 2006, 2013; Ross 1986) or as an affect- or emotion-based theory of trust (Lahno 2002; Faulkner 2015). The normative theory represents a profound transformation in how trust is understood, making substantial revisions to the epistemic perspective by avoiding its abstractions. However, normative accounts suffer from other one-sidedness. Their greatest weakness is that, in most of their versions, they fail to place the epistemic dimension of trust in a coherent relationship with its normative dimension (Kaminski 2017; Wiengarn 2021).

Yet the phenomenon of trust itself demands a conceptual framework that captures the unity of these dimensions. A crucial indication of this emerges from the experience of disappointed trust. When our trust is disappointed, it evokes two responses: (1) How could I have been so mistaken about you? and (2) How could you have deceived me like that? In short, we respond by questioning our own judgment (What could I have done better? How can I learn to grant trust more judiciously?) and, at the same time, by blaming the other person (You shouldn't have done that!). Here, reflection and learning on the one hand, and the moral gravity of betrayal on the other, converge in the way we may respond. This raises the question of what a theory of trust capable of grasping this unity might look like. Normative theories can only account for the normative response. They miss half of the phenomenon.[10]

The contours of such an approach, one that unifies our responses and captures trust in a non-reductive way, have already been implicitly drawn in references to the virtues of trustworthiness and trust. This is a third, alternative perspective (Kaminski 2020a, 2023), one that does not adopt a dualistic view of trust from the outset. Ascribing a virtue to someone presupposes that we have observed their behavior in ways that justify attributing that virtue to them. We rely on epistemic reasons grounded in how they act. Yet these are only epistemic reasons for attributing a virtue if the person in question does not stand in a merely external relationship to that virtue – say, behaving 'as if' being virtuous by chance or for strategic purposes alone. Instead, the person must incorporate a normative relationship to the virtue in question and recognize it as a virtue. The same is true of the person who ascribes a virtue: In order to ascribe a virtue rather than merely calling someone or something a reliable mechanism of future behavior, that person must also recognize the relevant trait as a virtue, thereby adopting a normative stance toward it.

Virtues and the Capacity Approach

Within the virtue-ethical framework, the epistemic dimension of trust finds its proper place. This resolves the important theoretical challenge of how to understand both dimensions of trust in a unified way. It does not, however, solve the practical challenge of determining whether a person – or, in our context, a technical system – is trustworthy. This challenge is especially daunting because it is difficult to distinguish between behavior that appears virtuous but is actually driven by strategic motives (for example, someone who is friendly only so long as the boss is present) and behavior that is genuinely motivated by an orientation toward virtue. Typically, familiarity can make this easier: You notice, for instance, that as soon as the boss leaves the room, your colleague's recently displayed friendliness evaporates.

---

10 At times, the literature tends to conflate distinct aspects. Alvarado (2023), for instance, appears to conflate the object of trust (information) with the type of trust (epistemic trust) when he argues that, since the primary function of AI is to serve as an epistemic enhancer, trust in AI is therefore epistemic in nature. But even if the latter were true – i.e., even if the sole function of AI were to provide information (which it is not, given that such systems are also used in decision-making contexts) – it would by no means follow that the manner in which AI is trusted must be exclusively epistemic. The sincerity of another person, for instance, is not trusted merely in an epistemic sense; the same holds for promises concerning the reliability of an AI system – they, too, need not be trusted solely on epistemic grounds.

Ordinarily, becoming familiar with a technology is one of the central aims of its design and a natural outcome of practical use (Kaminski 2010: Part II). However, the rapid pace of technological development, particularly in AI, together with the opacity of models and massive marketing efforts, significantly complicates this process. If we follow the argument set out above, the capacity approach would need to focus on examining how and to what extent the virtues relevant to the specific model in its context (justice, privacy, reliability, and so on) are realized by these systems, and also whether the design and operation of these systems are in fact guided by an orientation toward those virtues.

Clearly, this task overwhelms individual users. Often, there is not even a theory-independent way to explain what fair decisions by AI models might involve. Consequently, the capacity approach can only be understood as an institutional and collective endeavor. In turn, a well-grounded decision about whether to trust an AI system presupposes that the institutions, media outlets, and communities evaluating that system are themselves worthy of trust. Each individual's capacities are bound up with the capacities of the broader network. Hence, justified trust in AI systems requires justified trust in these networks. For this reason, AI ethics ultimately is AI politics. Hence trustworthy AI depends on digital governance in a broader social network.

References

AI HLEG (2019): Ethics guidelines for trustworthy AI. https://data.europa.eu/doi/10.2759/346720, last access: July 25, 2025.

Alvarado, Ramón (2023): "What Kind of Trust Does AI Deserve, If Any?", in: AI and Ethics 3(4), pp. 1169–1183.

Angius, Nicola and Plebe, Alessio (2023): "From Coding To Curing. Functions, Implementations, and Correctness in Deep Learning", in: Philosophy & Technology, 36(3), pp. 36–47.

Aristoteles (1994): Metaphysik, Neuausg. Reinbek b.H.: Rowohlt.

Aristoteles (2011): Nikomachische Ethik, 3rd ed., Wolf, Ursula (ed.), Reinbek b.H.: Rowohlt.

Arntzen, Friedrich (2007): Psychologie der Zeugenaussage. System der Glaubhaftigkeitsmerkmale, München: Beck.

Baier, Anette C. (1986): "Trust and Antitrust", in: Ethics 96(2), pp. 231–260.

Coady, Cecil (1992): Testimony. A philosophical study. Oxford: Oxford University Press.

Coleman, James S. (1990): Foundations of Social Theory, 1st ed., Cambridge, MA: Harvard University Press.

Dasgupta, Partha (1988): "Trust as a Commodity", in: Diego Gambetta (ed.), Trust. Making and Breaking Cooperative Relations, New York, NY: Basil Blackwell, pp. 49–72.

Durán, Juan M. and Formanek, Nico (2018): "Grounds for Trust. Essential Epistemic Opacity and Computational Reliabilism", in: Minds and Machines 28(4), pp. 645–666.

Ekman, Paul (2001): Telling Lies. Clues to Deceit in the Marketplace, Politics, and Marriage, New York: Norton & Company.

Ekman, Paul (2003): Emotions Revealed. Recognizing Faces and Feelings to Improve Communication and Emotional life, New York, NY: Holt & Company.

Ekman, Paul and Rosenberg, Erika L. (eds.) (2005): What the Face Reveals. Basic and Applied Studies of Spontaneous Expression Using the Facial Action Coding System (FACS), Oxford: Oxford University Press.

Faulkner, Paul (2007): "On Telling and Trusting", in: Mind 116(464), pp. 875–902.

Faulkner, Paul (2015): Knowledge on Trust, Oxford: Oxford University Press.

Floridi, Luciano (2019): "Establishing the Rules for Building Trustworthy AI", in: Nature Machine Intelligence 1(6), pp. 261–262.

Frevert, Ute (2013): Vertrauensfragen. Eine Obsession der Moderne. München: Beck.

Gambetta, Diego (ed.) (1988): "Can We Trust Trust?", in: Id. (ed.): Trust. Making and Breaking Cooperative Relations, New York, NY: Basil Blackwell, pp. 213–237.

Hardin, Russell (2002): Trust and Trustworthiness. New York: Russell Sage Foundation.

Hardin, Russell (2006): Trust. Cambridge: Polity.

Hartmann, Martin (2001): "Einleitung", in: Hartmann, Martin and Offe, Claus (eds.), Vertrauen. Die Grundlage des sozialen Zusammenhalts. Frankfurt/Main: Campus, pp. 7–34.

Hume, David. (2007): An Enquiry Concerning Human Understanding, Oxford: Oxford University Press.

Kaminski, Andreas (2010): Technik als Erwartung. Grundzüge einer allgemeinen Technikphilosophie, Bielefeld: Transcript.

Kaminski, Andreas (2014): "Lernende Maschinen: naturalisiert, transklassisch, nichttrivial? Ein Analysemodell ihrer informellen Wirkungsweise", in: Kaminski, Andreas and Gelhard, Andreas (eds.), Zur Philosophie der informellen Technisierung. Darmstadt: Wissenschaftliche Buchgesellschaft, pp. 58–81.

Kaminski, Andreas (2017): "Hat Vertrauen Gründe oder ist Vertrauen ein Grund? Eine (dialektische) Tugendtheorie von Vertrauen und Vertrauenswürdigkeit", in: Kertscher, Jens and Müller, Jan (eds.), Praxis und "zweite Natur" – Begründungsfiguren normativer Wirklichkeit in der Diskussion, Münster: Mentis, pp. 121–139.

Kaminski, Andreas (2020a): Die verwickelte Einfachheit von Vertrauen – und seine spekulative Struktur. Accepted habilitation thesis, Marburg.

Kaminski, A. (2020b): "Gründe geben. Maschinelles Lernen als Problem der Moralfähigkeit von Entscheidungen", in: Wiegerling, Klaus, Nerurkar, Michael and Wadephul, Christian (eds.), Datafizierung und Big Data. Ethische, anthropologische und wissenschaftstheoretische Perspektiven, Wiesbaden: Springer Fachmedien, pp. 151–174.

Kaminski, Andreas (2023): "Die Erfahrung gebrochenen Vertrauens", in: Zeitschrift für Kulturphilosophie 2023(2), pp. 96–115.

Lagerspetz, Olli (1998): Trust. The Tacit Demand, Dordrecht: Springer.

Lahno, Bernd (2002): Der Begriff des Vertrauens, Paderborn: Mentis.

Lauer, Josh (2017): Creditworthy. A History of Consumer Surveillance and Financial Identity in America, New York: Columbia University Press.

Moran, Richard (2005): "Problems of Sincerity", in: Proceedings of the Aristotelian Society 105(1), pp. 325–345.

Moran, Richard (2006): "Getting Told and Being Believed", in: Lackey, Jennifer and Sosa, Ernest (eds.), The Epistemology of Testimony, pp. 272–306.

Moran, Richard (2013): "Testimony, Illocution and the Second Person", in: Aristotelian Society Supplementary 87(1), pp. 115–135.

Nickel, Philip J., Franssen, Maarten and Kroes, Peter (2010): "Can We Make Sense of the Notion of Trustworthy Technology?", in: Knowledge, Technology & Policy 23(3), pp. 429–444.

Ross, Angus (1986): "Why Do We Believe What We Are Told?", in: Ratio 1, pp. 69–88.

Sporer, Siegfried L. (1997): "Realitätsüberwachungskriterien und forensische Glaubwürdigkeitskriterien im Vergleich. Validitätsüberprüfung anhand selbsterlebter und erfundener Geschichten", in: Greuel, Luise, Thomas, Fabian and Stadler, Michael (eds.), Psychologie der Zeugenaussage. Ergebnisse der rechtspsychologischen Forschung, Weinheim: Beltz, pp. 71–85.

Stern, William (ed.) (1903): Beiträge zur Psychologie der Aussage. Mit besonderer Berücksichtigung von Problemen der Rechtspflege, Pädagogik, Psychiatrie und Geschichtsforschung, Leipzig.

Thiebes, Scott, Lins, Sebastian and Sunyaev, Ali (2021): "Trustworthy artificial intelligence", in: Electronic Markets 31(2), pp. 447–464.

Wagner, Gerald (1994): "Vertrauen in Technik", in: Zeitschrift für Soziologie 23(2), pp. 145–157.

Wiengarn, Jörn (2021): Die Grammatik des Vertrauens. Eine Untersuchung in interpersoneller und epistemischer Hinsicht, Köln: transcript.

Explainable AI as a Component of Building Trust

The Case of Regulating Creditscoring

Katja Langenbucher

Abstract *The paper takes up the notions of trust and explainability in the GDPR and in upcoming German legislation, using AI-based credit scoring as an illustration. It offers an overview of methods of explainable AI, stressing differences between computer scientists, legal scholars, and legislators. Counterfactual explainability, the paper claims, might be useful along the lines of the ECJ decision (Court of Justice of the European Union 2025).*

> "The purpose of this Regulation is to improve the functioning of the internal market by laying down a uniform legal framework in particular for the development [...] and the use of artificial intelligence systems (AI systems) in the Union in accordance with Union values to promote the uptake of human centric and trustworthy artificial intelligence". This is how the first recital of EU Regulation 2024/1689 of 13 June 2024 (AI Act) on harmonized rules concerning artificial intelligence (AI) starts. At what point an AI system counts as trustworthy is not defined in the AI Act. Instead, the term appears in the law in a variety of contexts. We find, for instance, the uptake of "human centric and trustworthy" AI (Recital 1, AI Act), the goal to develop "secure, trustworthy and ethical AI" (Recital 8, AI Act), along with "accuracy, reliability and transparency [...] to avoid adverse impacts, retain public trust and ensure accountability and effective redress" (Recital 59, AI Act).

1. The Concept of Trust in the AI Act

What counts as "trustworthy" varies significantly across disciplines and context (see Kaminski 2025 on philosophy; Zhang et al. 2024 on psychology; Aljohani et al. 2025 on medicine; Breuer and McDermott 2008 on economics). The AI Act does not define the concept of trust or of trustworthiness. Instead, it mostly appears as an element of explaining the EU Commission's regulatory philosophy, based on everyday language.

Trust. Arguably, one of the first times the term "trust" surfaces in the context of AI regulation is in the 2018 EU Commission Communication "Artificial Intelligence for Europe" (EU Commission 2018). That strategy references the GDPR as a "*major step for building trust, essential in the long-term for both, people and companies*", along with the – then – proposals for the flow of non-personal data, the e-Privacy Regulation and the Cybersecurity Act. Additionally, the Communication emphasizes the role of private rights of actions if things go wrong: "*A high level of safety and an efficient redress mechanism for victims in case of damages helps to build user trust and social acceptance of these technologies*".

Trustworthiness. A year later, the 2019 High Level Expert Group Ethics Guidelines for Trustworthy AI, endorsed by the EU Commission (EU Commission 2019), introduced "*trustworthy AI*" as a guiding concept. Trustworthy AI "*should be (1) lawful – respecting all applicable laws and regulations, (2) ethical – respecting ethical principles and values, (3) robust – both from a technical perspective while taking into account its social environment*".

An ecosystem of trust. In its 2020 White Paper on AI (EU Commission 2020), the Commission further developed the concept into one of two pillars of its AI Regulation. The first one is an "*ecosystem of excellence*", the second an "*ecosystem of trust*". The latter is "*a policy objective in itself*" and "*should give citizens the confidence to take up AI applications and give companies and public organizations the legal certainty to innovate using AI*". Along those lines, the AI Act uses the term as a goal, justifying the Act's risk-based approach: "*To ensure a high level of trustworthiness, certain mandatory requirements should apply to high-risk AI systems*" (Recital 64, AI Act). "*While the risk-based approach is the basis for a proportionate and effective set of binding rules*", Recital 27 stipulates, "*it is important to recall the 2019 Ethics guidelines for trustworthy AI developed by the independent AI HLEG appointed by the Commission. In those guidelines the AI HLEG developed seven non-binding ethical principles for AI [...]. These seven principles include human agency and oversight; technical robustness and safety; privacy and data governance; transparency; diversity, non-discrimination and fairness; societal and environmental well-being and accountability*".

Transparency. One of the seven principles of trustworthy AI is transparency. "Transparency", we find in that same Recital 27, "means that AI systems are developed and used in a way that allows appropriate traceability and explainability". Traceability targets compliance, documentation and performance during the lifetime of an AI system (Recitals 27, 53, 71, Art. 12(2)). Explainability, by contrast, references – like in 2018 – the link between trust and private rights of action. The "exercise of important procedural fundamental rights, such as the right to an effective remedy and to a fair trial as well as the right of defense and the presumption of innocence" requires the citizen to have appropriate information to back up a litigation claim. Faced with AI, this can be challenging, if the potential litigant has no understanding of what triggered a particular decision and who might be responsible for it.

Explainability. Against this background, explainability is a core component of trust-based AI regulation. This is not to be understood as a *necessary* element. Arguably, in most low-risk use cases, consumers do not care about an explanation of how the AI produced its result. Picture-generating models provide an illustration: Using an AI to design a birthday card or enhance a power point presentation does not call for an explanation of how the AI did it, as long as the user enjoys the picture. This is different for high-risk use cases. A person who receives a lower credit score than he or she expected is likely to demand explainability of what led to the score, to change behavior or to prepare litigation.

2. Explainability Rights

Art. 22, 15 GDPR provide core private rights, with the AI Act and (for credit-scoring) the Consumer Credit Directive adding finishing touches, as it were. Art. 22 GDPR regulates instances of automated decision-making that produce "*legal effects* [...] *or similarly significantly*" affect the data subject. In line with the Regulation's general approach, the rule starts with a prohibition of this type of automated decision-making, Art. 22(1) GDPR. Then, Art. 22(2), (3) GDPR follow up with exceptions to the ground rule.

Explainability of automated decision-making is covered in Art. 15(1)(h) GDPR. The rule provides the data subject with a right to "*meaningful information about the logic involved*". While the legislator might not have had AI-based decision-making in mind, the text of the Regulation covers it, and two recent ECJ-decisions, both concerning credit-scoring, have broadened, rather than narrowed, its scope.

Against this background, it is unsurprising that it was late in the process of passing the AI Act, that EU legislators decided to, in addition, enshrine a right to an explanation of individual decision-making in Art. 86(1): "*Any affected person subject to a decision which is taken by the deployer on the basis of the output from a high-risk AI system* [...] *which produces legal effects or similarly significantly affects that person in a way that they consider to have an adverse impact on their health, safety or fundamental rights shall have the right to obtain from the deployer clear and meaningful explanations of the role of the AI system in the decision-making procedure*".

The text seems to mostly repeat Art. 15 GDPR. Both rules concern rights to receive an explanation, and Art. 86(3) AI Act mentions that its para. (1) shall not apply if the right it confers is otherwise provided for under Union law. Automated decision-making, including profiling, triggers Art. 15(1)(h) GDPR. Rather vaguely, it then speaks about access to information about *the logic involved*. Not each instance of automated decision-making involves an "AI system" under Art. 3(1) AI Act, which requires "*varying levels of autonomy*" and inferences from the input the system receives on *how to generate outputs such as predictions, content, recommendations, or decisions*. In that sense, the scope of Art. 86(1) AI Act is narrower, because it applies only to AI systems. There is a slight variation in the text, if compared to Art. 15(1)(h) GDPR: The AI Act concerns a right to obtain an explanation on "*the role of the AI system in the decision-making procedure*" (*sur le role du système d'IA dans la procedure décisionelle; zur Rolle des KI-Systems im Entscheidungsprozess; sul ruolo del sistema di IA nella procedura decisionale*). The GDPR, by contrast, focuses on the *logic involved* in the automated decision-making, including profiling. One might read this as the AI Act being more interested in an understanding of what the AI system contributes to the overall decision-making process, whereas the GDPR has the functioning of the AI itself in mind. However, the wording is similar and, arguably, Art. 86(1) AI Act will mostly be relevant to fill gaps Art. 15(1)(h) GDPR might leave.

3. An illustration: AI-based Credit Scoring

Like the concept of "trust", the concept of "explainability" varies according to context. This makes an inductive approach useful, presenting one example for the role of explainability in the general context of trustworthy AI. AI-based credit scoring is an apt illustration: It qualifies as a high-risk AI-system under Art. 6(2), Annex III No. 5 AI Act. Building a credit score involves large amounts of data, which brings the GDPR into play. Two recent European Court of Justice decisions have found that computing a credit score counts as automated decision-making under Art. 22 GDPR, hence, Art. 15(1)(h) GDPR's right to be informed about the "logic involved" applies. Additionally, there is sectoral EU legislation concerning creditors and German legislation in the making on credit scoring.

Which role does explainability play in the context of credit scoring? A consumer might, for several reasons, ask for an "explanation": To verify the accuracy of the data used, to decide whether he has given his consent for data use, to adapt his behavior, in the hope of receiving a better score in the future, or to litigate, should the score seem unfairly low. In that way, an "explainable" credit score can be conducive to building trust.

The Consumer Credit Directive. Directive (EU) 2023/2225 (CCD) provides a sectoral private right of action, complementing the GDPR. It is different from the GDPR in that its scope extends solely to the relationship between creditor and borrower. Scoring agencies fall outside the CCD. It starts from the assumption that "*artificial intelligence (AI) systems can be easily deployed in multiple sectors of the economy and society*" (Recital 46, CCD). Following up on the GDPR, the CCD explains that "*the consumer should have the right to obtain a meaningful, comprehensive explanation of the assessment made and of the functioning of the automated processing used, including the main variables, the logic and risks involved, as well as the right to express the consumer's point of view and to request a review of the assessment of the creditworthiness and a review of the decision on whether to grant credit*" (Recital 56, CCD). Art. 18(8) CCD lays down the details: "*where the creditworthiness assessment involves the use of automated processing of personal data, Member*

States shall ensure that the consumer has the right to [...]: (a) request and obtain from the creditor human intervention, consisting of the right to request and obtain from the creditor a clear and comprehensible explanation of the assessment of creditworthiness, including on the logic and risks involved in the automated process of personal data as well as its significance and effects on the decision".

The German Bundesdatenschutzgesetz-draft. In Germany, legislators have been considering private rights of action under national law to further support the consumer in a credit underwriting situation. To this end, a law on credit scoring (drafted before the new government started in 2025, BT-Drucksache 20/10859) takes advantage of the discretion that the GDPR leaves for Member States. § 37a *Bundesdatenschutzgesetz*-draft, first, references Art. 22 GDPR which allows automated decision-making, such as credit scoring, if Member State law lays down the details. Second, § 37a *Bundesdatenschutzgesetz*-draft requires an "input control". It specifies certain data points, for instance, sensitive data under Art. 9 GDPR, that may not be used by a credit scoring company. The question whether an input control is a smart form of legislating is beyond the scope of this paper. Suffice to say that, given inferences that can be drawn from other data points (the problem of redundant encoding, see Barocas and Selbst 2016), this strategy will only help if an AI system works with a very limited set of data points (see Solove 2024 for the argument, that the concept of sensitive data is at a dead end). Third, the German draft law stipulates a form of quality control. The data used must be processed "*on the basis of an appropriate mathematical-statistical procedure which is demonstrably relevant to compute the probability of a specific conduct*".

Fourth and last, the rule includes its own version of a right to transparency. The company, which produces a credit score, must deliver information "*in a* precise, *transparent, understandable and easily accessible form as well as in clear and simple language*". Four elements must be explained: "*the personal data used, the weight of data points that influence the score most importantly, the meaning of the specific score and the score itself*". The law, which has not yet entered into force and may be changed under the new government, includes detailed explanations of what it has in mind as to transparency rights. More specifically, it requires scoring companies to use language that is targeted to its audience and to reflect upon its "cognitive capabilities".

4. "Explaining AI" – A short overview

Providing an explanation of automated decision-making is clearly at the forefront of the legislative endeavors mentioned in the previous paragraph. This (mostly correctly) presumes that those affected from automated decision-making have a general interest to understand how the decision was made (but see Langenbucher 2024 on "black-box rights"). However, the details of a transparency-enhancing private right of action depend very much on context. Understanding why a doctor suggests staying calm, after his AI predicted a low probability that a beauty mark is cancerous, raises very different issues than following up on an AI-based credit score or evaluating an AI-supported judicial decision to let someone go free on bail. In a low-risk environment, understanding an AI's inner workings might not be relevant and might not justify the drop in predictive performance that is often associated with explainable AI (Molnar 2022: 3.1.). By contrast, the necessities of safety measures and testing, the detection of bias or the wish to increase social acceptance might call for the use of explainable systems (ibid.: 3.1.).

Against this background, it is tempting to draw on computer science efforts to provide "explainable" AI (XAI). However, it is important to bear in mind that, as stressed above, context matters. A computer scientist who employs an XAI model will be interested in different questions than a consumer looking at his score, a lawyer preparing an anti-discrimination lawsuit, or a banking supervisor who runs a risk-management check. Some of this has to do with varying competences and capabilities of the actors (Kaminski 2025). Additionally, their specific reasons for requiring an explanation determine what is useful for them. The computer scientist will wish to gain a better understanding of the steps the AI takes, for instance, across the different layers of a neural network. Whether the outcome adequately represents an individual's capability to pay back a loan is not the computer scientist's concern, especially, if possible flaws have nothing to do with the model, but go back to faulty data. By contrast, neither the consumer nor the lawyer are overly interested in the inner workings of the model. Their core interest will usually lie with the data used and the predictive power of the score. The banking supervisor's interest is situated somewhere in between. Data that produce inappropriate uncertainty as to adequate representation of a portfolio of creditors may not be used. The same goes for models that are inappropriate for that purpose.

Explainability is not the same as substantive control. I might very well understand how a decision was made but still consider it unfair or unlawful. Often, explainability tells us something about the procedure of decision-making. This might be done, for instance, by reproducing each step, by highlighting core elements, or by producing counterfactuals. When discussing explanations in the context of AI, it is helpful to distinguish between two approaches: Using models that are inherently interpretable due to their simpler structure (like linear regression, decision trees, or k-nearest neighbors) and using post-hoc explanatory techniques designed to shed light on the behavior of more complex, often opaque black-box models (like deep neural networks). The field of XAI is primarily concerned with the latter, developing methods like LIME, SHAP, or DiCE (see Dubovitskaya and Bosold 2025). Efforts of credit scoring companies, such as the German SCHUFA, rely on the former in explaining consumers the impact of individual components of their credit score.

What all these situations have in common is the legislator's assumption that the contribution of each feature a model uses can be computed and disclosed. For linear regression models, this is correct (Molnar 2022: 9.5.1.). For more complex, non-linear models, statisticians have produced a wide range of potential explanations that vary in usefulness according to context (overview at Freiesleben 2022; pointing to the irony that XAI models add a second layer of complex models on the blackbox model: Nisevic, Cuypers and De Bruyne 2025 noting an "XAI chaos").

Local surrogate methods. One common approach of XAI are local surrogate methods (Dubovitskaya and Bosold 2025). These methods produce a local explanation of the AI decision. An explanation is local, rather than global, if it targets the region around the prediction of interest. What it produces does not attempt to explain the inner workings of the entire model, but only one specific prediction, in our case: the credit score for a specific customer.

These methods are called surrogate, because they involve creating a simpler, interpretable model that approximates the behavior of a more complex model in the local region. The first step is to create a set of perturbed samples around the prediction point. This includes varying all features randomly within a local neighborhood. The samples are weighted according to their distance from the original point, with closer samples receiving higher weights to ensure that the surrogate model focuses on the immediate vicinity of the prediction. For these samples, the original, complex model is run to predict outcomes. Then, one trains a simpler, inherently interpretable 'surrogate' model on these perturbed samples and their predictions, aiming to mimic the complex model's behavior locally. This produces a local, surrogate explanation of the black-box model. A common choice for this surrogate model is a decision tree because these are relatively easy to understand. One takes the set of perturbed samples and their corresponding predictions from the complex model and trains a simple decision tree model using this local data. This trained decision tree then serves as the local explanation.

Given that a surrogate model does not try to explain the computation of the complex model, it is model-agnostic. This means it can (locally) explain predictions from any complex model. Taking these characteristics together, the technique is often called LIME (Local Interpretable Model-agnostic Explanations; Ribeiro, Singh and Guestrin 2016).

Advantages of local surrogate models, apart from being model-agnostic, are that they are simple to understand and capture the model's behavior around the specific point of interest, which might be different from its global behavior. It takes a shot at highlighting feature importance, something that, arguably, § 37a *Bundesdatenschutzgesetz*-draft was looking for, when it asked for "weights". A decision-tree algorithm determines, for instance, which features are most informative for splitting the data to match the complex model's predictions. If changing a feature significantly alters the complex model's predictions, for instance, the number of credit cards or open bills, the decision tree is likely to use that feature in its top-level splits. To make sure that the feature importance isn't skewed, one would perturb all features uniformly, vary the perturbation strategy, and do multiple runs. Still, a model like LIME assumes linear behavior of the model locally. It is unclear whether there is a solid theoretical basis for this assumption (doubting this: Molnar 2022: 9.5.4.).

Shapley additive explanations. Another powerful method to explain a complex model's predictions is SHAP (SHapley Additive exPlanations). This method is based on cooperative game theory. SHAP assumes that each feature value is a player in a game where the prediction is the payout. Shapley values in cooperative games demonstrate how to fairly distribute that profit among all players. Used to explain an ML-prediction, the first step is to single out players: Each individual feature the model uses counts as one player. Second, the prediction becomes the game's "payout" (ibid.: 9.5.1.). In this way, SHAP calculates the contribution of each feature to the actual prediction that the model arrived at by systematically including and excluding features to simulate different scenarios.

SHAP starts with a baseline prediction: the average model output over the entire dataset, for instance, an average score. Then, SHAP calculates how each feature, such as the number of credit cards a consumer holds, pushes the prediction away from the baseline. To do this, one generates perturbed samples (see ibid.: 9.5.5.) where each sample represents a different combination of features being "present" or "absent". One then inputs each sample into the complex model and records each output prediction. These output predictions help to understand how the (original, complex) model behaves in the local neighborhood around the relevant situation. By comparing how the prediction changes as features are perturbed, SHAP can deduce the importance of each individual feature. The predictions for the perturbed samples are compared against the prediction for the original sample and the marginal contribution of each feature is calculated. This process is repeated in various combinations of features. The Shapley value is the average marginal contribution of one feature across all possible combinations of features. It can range from one single feature to all features in the model. Additionally, as we will see further below, SHAP can also produce global explanations.

SHAP is different from LIME in that it uses the original, complex model. In this way, it offers the potential for global interpretability of outputs by aggregating SHAP values across many predictions. Note that this is still a second-hand explanation, as it were. SHAP does not identify the way in which features move through the layers of a neural network. Even less does it identify real causal relations between data points.

Additionally, SHAP allows to identify feature interactions, for instance: How does having two credit cards impact the number of open bills. SHAP calculates this by comparing the effect of all features together against their individual and pairwise effects. Furthermore, SHAP but not LIME, achieves a fair distribution of each importance, because it considers all possible coalitions, calculates the marginal contribution of each feature across these coalitions, and averages these contributions. LIME, by contrast, focuses on local approximations around specific predictions and may fail to capture the true importance of features in the global context of the model.

Applicability in a legal context. What might a use case for these models in a legal context look like? Above, we used the example of a banking supervisor or a consumer advocacy group's interest to receive a detailed explanation of specific feature values. SHAP allows for that. Assume a consumer advocacy group is weighing the odds of a lawsuit based on indirect gender discrimination. One of the test prongs will be to show that the loan applicants are "similarly situated" – you cannot compare apples with pears. The consumer advocacy group might argue as follows: Let's have a look at the subgroup of university professors in the highest income bracket applying for a loan. They would like to know the relative importance of the gender of a university professor in that subgroup. SHAP will give you this value, whereas local models do not allow for contrastive explanations (ibid.: 9.5.4.).

Drawbacks of SHAP. Depending on context, it is important to note that SHAP requires a representative background dataset to avoid unrealistic feature-value combinations. Another disadvantage can be that many versions of SHAP assume feature independence (ibid.: 9.5.5.), which is often unrealistic in an empirical setting. If that assumption breaks down, the model's explanation is less (or not at all) useful. Let us follow up on the consumer advocacy group-example: Many features of individual loan applicants in the subgroup of university professors will be correlated (for the assumption that everything correlates with race, see: Prince and Schwarcz 2020; Langenbucher 2022: 22–27). In fact, the legal doctrine of indirect discrimination was developed to cope with correlated features: If an employer discriminates on the basis of part-time work, he does not directly discriminate against women. However, if the percentage of part-time workers is predominantly female, he discriminates indirectly because gender and part-time work correlate narrowly. Such correlations can influence a model's prediction if two features are highly correlated or if one specific feature is slightly correlated with many features. For SHAP, this can raise important challenges, especially if a truly model-agnostic SHAP is used. Some versions of SHAP cope better than others. TreeSHAP is optimized for tree ensembles like random forest and gradient boosting machines and can better account for feature independence than, for instance, KernelSHAP, even though some extensions to this latter method are being proposed (Shuyang 2024; on yet another approach, Generalized DeepSHAP, see Chen, Lundberg and Lee 2022).

Contrastive and counterfactual explanations. SHAP gives what computer scientists call a "contrastive" explanation. It shows why one specific prediction differs from a baseline, for instance: compared to the average credit score, having seven credit cards lowers the score by 5 points. Additionally, SHAP can compare one prediction to a subset of the data set or even to one single instance, by calculating differences in their feature contributions, for example, scores above or below the threshold a bank sets having to do with income or open bills. This form of transparent, quantitative feature attribution is one of the strengths of SHAP. It answers the question of why a certain outcome (e.g., a score of 10) was reached, instead of a different outcome (e.g., a score of 8). If a user is not so much interested in comparing predictions across instances, but on receiving a recommendation (e.g., get rid of two of your eight credit cards), a counterfactual explanation might be useful. Counterfactual explanations tell us what features must change to produce a certain outcome (e.g., a score) by describing the smallest change to the feature value (e.g., the number of credit cards) that changes the prediction to a predefined output (e.g., a certain threshold score). In that way, counterfactual explanations answer "what if?" questions (Wachter, Mittelstadt and Russell 2018). The goal is to find a set of examples that not only achieve the desired outcome (validity) but are also as close as possible to the original data point (proximity) and differ significantly from each other (diversity) to represent various actionable paths (Mothilal, Sharma and Tan 2020). There are both model-agnostic and model-specific counterfactual explanation methods (Molnar 2022: 15). One explanatory model that delivers counterfactual explanations is DiCE (Mothilal, Sharma and Tan 2020; de Oliveira, Sörensen and Martens 2024). While LIME and SHAP primarily focus on highlighting the importance of individual features for a specific prediction, DiCE proactively generates multiple diverse counterfactual examples to show different ways the outcome could be changed (ibid.; Jain, Sangroya and Vig 2025; Dominici et al. 2025).

Drawbacks of counterfactual models. At first glance, counterfactual explanations, like those generated by DiCE, seem to provide an ideal tool for explaining credit scoring. The potential borrower learns how he can "play" with relevant features, focusing on a small number of changes. However, they generally suffer from several issues concerning their robustness. Minor changes to the underlying model can invalidate the previously generated explanation (Upadhyay, Joshi and Lakkaraju 2021; Hamman et al. 2023). Similarly, slight variations in the input data can lead to entirely different counterfactual suggestions (Slack et al. 2021; Artelt et al. 2021). While methods to enhance robustness exist, they often come with significantly increased computational costs (Jiang et al. 2024). Furthermore, there is no guarantee that the examples generated by DiCE are always realistic or plausible within the data's context or actually feasible for the user to implement (Salimi et al. 2023; Barr et al. 2021), e.g., change your age or gender to receive a better score. Lastly, counterfactual explanations suffer from what computer scientists call the "Rashomon effect", namely that there exist many equally good predictive models for the same dataset (Rudin et al. 2024). Applied to counterfactual models, this translates as receiving multiple different counterfactual explanations, each telling a different story and, possibly, contradicting each other.

Dandl et al. 2020 provide the following example, illustrating their multi-objective counterfactual model (MOC) in the context of credit-scoring:

Figure 1: Baseline data of a sample consumer for counterfactual modelling.[1]

Age	Sex	Job	Housing	Saving accounts	Checking account	Credit amount	Duration	Purpose
22	Female	2	Own	Little	Moderate	5951	48	Radio/TV

Their model generated a total of 136 counterfactuals and then focused on 82 of them with predictions within [0.5,1]. They produced a response surface plot, suggesting decreasing credit duration and credit amount:

Figure 2: Example for a Response surface plot generated from a counterfactual model.[2]

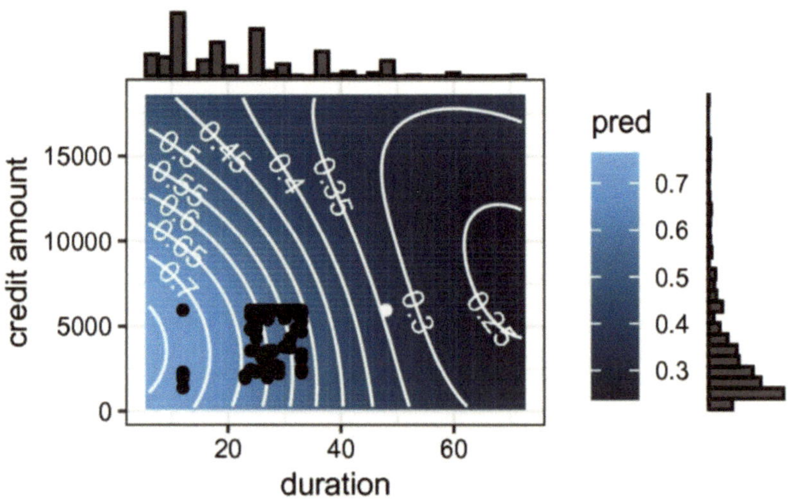

(b) Response surface plot

Molnar (2022: 15) provides a variation on this example, based on the dataset used in Dandl et al. 2020. The consumer is described as follows:

1 Illustration by Dandl (et al 2020) of their multi-objective counterfactual model (MOC) in the context of credit-scoring. Screenshot from 13.10.2025; used with the authors' consent.

2 Response surface plot by Dandl (et al 2020), suggesting decreasing credit duration and credit amount. Screenshot from 13.10.2025; used with the authors' consent.

Figure 3: Baseline data for a variation of the first credit scoring example.[3]

Table 15.1: Feature values of a particular customer

age	sex	job	housing	savings	amount	dur.	purpose
58	f	unskilled	free	little	6143	48	car

The model predicts that the probability that the consumer gets her preferred score is 24.2%. Her interest is to employ a counterfactual explanatory model to understand what she needs to change as to her input features to reach a probability of >50% to get the preferred score. The model displays the following results:

Figure 4: Example counterfactual results illustrating the effects of the suggested changes on the predicted credit score.[4]

Table 15.2: The ten best counterfactuals found for the customer

age	sex	job	amount	dur.	o_2	o_3	o_4	f(x')
		skilled		-20	0.108	2	0.036	0.501
		skilled		-24	0.114	2	0.029	0.525
		skilled		-22	0.111	2	0.033	0.513
-6		skilled		-24	0.126	3	0.018	0.505
-3		skilled		-24	0.120	3	0.024	0.515
-1		skilled		-24	0.116	3	0.027	0.522
-3	m			-24	0.195	3	0.012	0.501
-6	m			-25	0.202	3	0.011	0.501
-30	m	skilled		-24	0.285	4	0.005	0.590
-4	m		-1254	-24	0.204	4	0.002	0.506

3 Variation on the Dandl (et al 2020) example by Molnar (2022, 15), based on the dataset used in Dandl et al (2020). Screenshot from 13.10.2025; used with the author's consent.

4 Variation on the Dandl (et al 2020) example by Molnar (2022, 15), based on the dataset used in Dandl et al (2020). Screenshot from 13.10.2025; used with the author's consent.

Some of these help: The consumer learns, for instance, that she should lower the duration of the loan. Others are examples of complicated or even unrealistic suggestions: Seven of the ten best counterfactuals suggest to become "skilled", but it is unclear whether the potential borrower has that option. She cannot change her gender to "m" (as suggested by four of the ten best counterfactuals) or lower her age (as suggested by seven of the ten best counterfactuals).

5. Coming Full Circle: Credit Scoring and Explainability

The rough-and-ready overview of different explanatory strategies has highlighted how these work and what some of their advantages and disadvantages are. An important feature to keep in mind is the probabilistic nature of AI systems which accounts for accurate predictions without revealing the underlying causal mechanisms. Often, this produces a disconnect with the law's expectations. Hence, in a legal context, picking the best – or the second best – explanatory strategy depends very much on context.

If a bank asks its financial supervisor to allow it to use a certain model, global explanatory power will matter a lot. By contrast, if a consumer asks for a good-enough, easy-to-understand explanation of his credit score, while the profiler will want to keep his trade secrets, a counterfactual model might be sufficient. For the consumer, it will often be more important to understand his options for behavioral change when confronted with his score. To learn about those, a local explanation that approximates what the complex model does and limits itself to a sparse explanation will often suffice (see Molnar 2022: 9.5.5.).

European Court of Justice in Dun & Bradstreet (D&B). A recent decision by the European Court of Justice nicely illustrates the legal and practical relevance of providing these explanations to AI-based predictions – particularly counterfactual ones that are meaningful to the individual. The case (C-203/22, judgment Feb 27, 2025) involved a plaintiff who was denied a mobile phone contract based on an opaque credit score provided by D&B, who subsequently refused to disclose a detailed explanation of the underlying computation, citing trade secrets (Langenbucher and Bauer 2025). In interpreting the requirement of "meaningful information about the logic involved" under Art. 15(1)(h) GDPR, the ECJ clarified that a credit scoring company is not required to provide complex algorithms. Instead, the Court emphasized the need for "clear, understandable explanations", seen from an average consumer's point of view. Crucially, the Court suggested that explaining how changes to the individual's data would have led to a different score could satisfy this requirement – an approach strongly aligning with the concept of counterfactual explanations (ibid.). Methods like DiCE, designed to generate diverse, actionable counterfactuals (e.g., "If the consumer had one less credit card…"), thus present a potential technical solution for fulfilling these transparency obligations while respecting intellectual property rights (ibid.).

However, this raises further practical questions: While the ECJ endorsed the possibility of courts reviewing information, concerns remain about whether courts possess the necessary technical expertise to adequately assess the validity and potential manipulation of counterfactuals generated solely by the scoring entity. This challenge suggests a likely need for independent technical experts or neutral intermediaries to verify the reliability and completeness of such explanations in practice (ibid.).

XAI methods – from local approximations like LIME, game-theoretic approaches like SHAP, to counterfactual explanations using DiCE – offer various tools to make the functioning of AI systems more understandable. However, as the analysis of legal requirements from GDPR, the AI Act, and more specific regulations like the German *Bundesdatenschutzgesetz*-draft shows, explainability often serves as a vehicle to realize core aspects of the "trustworthiness" sought by the legislator – such as traceability, fairness, accountability, and the possibility of effective redress. The choice of the "right" explanation method depends heavily on context-specific needs: While a bank supervisor might wish to gain access to global insights (SHAP), an affected consumer might be looking for concrete options for action (DiCE, MOC). The challenge for the future lies in integrating these technical possibilities with the legal framework, combining usability in practice with risk-awareness for all parties concerned.

References

Aljohani, Manar et al. (2025): "A Comprehensive Survey on the Trustworthiness of Large Language Models in Healthcare", arXiv:2502.15871.

Artelt, André et al. (2021): "Evaluating Robustness of Counterfactual Explanations", in: 2021 IEEE Symposium Series on Computational Intelligence (SSCI), pp. 1–8, arXiv.2103.02354.

Barocas, Solon and Selbst, Andrew (2016): "Big Data's Disparate Impact", in: California Law Review 104(3), pp. 671–732.

Barr, Kyle et al. (2021): "Counterfactual Explanations via Latent Space Projection and Interpolation", arXiv:2112.00890.

Breuer, Janice and McDermott, John (2008): "Trustworthiness and Economic Performance", https://papers.ssrn.com/sol3/papers.cfm?abstract_id=1314844.

Chen, Hugh, Lundberg, Scott M. and Lee, Su-In (2022): "Explaining a Series of Models by Propagating Shapley Values", in: Nature Communications 13, 4512.

Dandl, Susanne et al. (2020): "Multi-Objective Counterfactual Explanations", https://link.springer.com/chapter/10.1007/978-3-030-58112-1_31.

Dominici, Gabriele et al. (2025): “Counterfactual Concept Bottleneck Models”, https://openreview.net/forum?id=w7pMjyjsKN.

Dubovitskaya, Elena and Bosold, Gregor (2025): “Right of Explanation of AI Decisions” [Beitrag in diesem Band].

Freiesleben, Timo (2022): “What Does Explainable AI Explain”, Dissertation, LMU Munich, https://edoc.ub.uni-muenchen.de/31933/1/Freiesleben_Timo.pdf.

Hamman, Matthew et al. (2023): “Robust Counterfactual Explanations for Neural Networks With Probabilistic Guarantees”, in: Proceedings of the 40th International Conference on Machine Learning (ICML 2023), PMLR 202, pp. 12384–12401.

Jain, Suparshva, Sangroya, Amit and Vig, Lovekesh (2025): “DifCluE: Generating Counterfactual Explanations with Diffusion Autoencoders and Modal Clustering”, in: Proceedings of ACM Conference, New York, USA, arXiv:2502.11509v1.

Jiang, Ziyi, Leofante, Francesco, Rago, Antonio and Toni, Francesca (2024): “Robust Counterfactual Explanations in Machine Learning. A Survey”, arXiv:2402.01928.

Kaminski, Andreas (2025): “Trust in AI. A Unified Approach” [Beitrag in diesem Band].

Langenbucher, Katja (2022): “Consumer Credit in The Age of AI. Beyond Anti-Discrimination Law”, ECGI Law Working Paper 663, https://papers.ssrn.com/sol3/papers.cfm?abstract_id=4298261.

Langenbucher, Katja (2024): “Financial Profiling”, Conference on Mapping and Governing the Online World; Ascona, Switzerland, https://www.researchgate.net/publication/381127142_Financial_Profiling.

Langenbucher, Katja and Bauer, Kevin (2025): “Explaining Credit Scores. The ECJ Rules on Automated Credit Assesments”, Compliance & Enforcement, PCCE at NYU, https://wp.nyu.edu/compliance_enforcement/2025/03/18/explaining-credit-scores-the-ecj-rules-on-automated-credit-assessments/, last access: June 18, 2025.

Molnar, Christoph (2022): Interpretable Machine Learning. A Guide for Making Black Box Models Explainable, https://christophm.github.io/interpretable-ml-book/.

Mothilal, Ramaravind K., Sharma, Amit and Tan, Chenhao (2020): “Explaining Machine Learning Classifiers Through Diverse Counterfactual Explanations”, in: Proceedings of the 2020 Conference on Fairness, Accountability, and Transparency (FAT* ‘20), Association for Computing Machinery, New York, NY, USA, pp. 607–617.

Nisevic, Maja, Cuypers, Arno and de Bruyne, Jan (2025): “Explainable AI. Can the AI Act and the GDPR go out for a Date?”, International Joint Conference on Neural Networks, Yokohama, Japan, https://papers.ssrn.com/sol3/papers.cfm?abstract_id=5056022.

de Oliveira, Raphael M. B. de, Sörensen, Kenneth and Martens, David (2024): "A Model-agnostic and Data-independent Tabu Search Algorithm to Generate Counterfactuals for Tabular, Image, and Text Data", European Journal of Operational Research 317(2), pp. 286–302.

Prince, Anya and Schwarcz, Daniel (2020): "Proxy Discrimination in the Age of Artificial Intelligence and Big Data", Iowa Law Review 105, pp. 1257–1318.

Ribeiro, Marco Tulio, Singh, Sameer and Guestrin, Carlos (2016): "Why Should I Trust You? Explaining the Predictions of any Classifier", in: Proceedings of the 22nd ACM SIGKDD International Conference on Knowledge Discovery and Data Mining, ACM, pp. 1135–1144, California, USA, arXiv:1602.04938.

Rudin, Cynthia et al. (2024): "Amazing Things Come From Having Many Good Models", arXiv:2407.04846v1.

Shuyang, Xiang (2024): "KernelSHAP Can Be Misleading With Correlated Predictors", TDS Archive, https://towardsdatascience.com/kernelshap-can-be-misleading-with-correlated-predictors-9f64108f7cfb/.

Salimi, Pedram et al. (2023): "Towards Feasible Counterfactual Explanations. A Taxonomy Guided Template-based NLG Method, Frontiers in Artificial Intelligence and Applications", 372, pp. 2057–2064.

Slack, Dylan et al. (2021): "Counterfactual Explanations Can Be Manipulated", in: Advances in 34 Neural Information Processing Systems (NeurIPS 2021), https://proceedings.neurips.cc/paper/2021/hash/009c434cab57de48a31f6b669e7ba266-Abstract.html.

Solove, Daniel J. (2024): "Data Is What Data Does. Regulating Based on Harm and Risk Instead of Sensitive Data", in: Northwestern University Law Review 118, pp. 1081–1138.

Upadhyay, Samanvay, Joshi, Shweta and Lakkaraju, Himabindu (2021): "Towards Robust and Reliable Algorithmic Recourse", in: Advances in Neural Information Processing Systems 34 (NeurIPS 2021), pp. 29536–29548.

Wachter, Sandra, Mittelstadt, Brent and Russell, Chris (2018): "Counterfactual Explanations Without Opening the Black Box. Automated Decisions and the GDPR", in: Harvard Journal of Law & Technology 31, pp. 841–887.

Zhang, Zhen et al. (2024): "Visual Analysis of Trustworthiness Studies. Based in the Web of Science Database", in: Front. Psychol 15, 1351425.

Legal Sources

Court of Justice of the European Union, First Chamber. Case C-203/22 Magistrat der Stadt Wien v Dun & Bradstreet Austria GmbH. Judgment of 27 February 2025, ECLI:EU:C:2025:117.

European Commission (2018): "Communication from the Commission, Artificial Intelligence for Europe", COM(2018) 237 final, https://eur-lex.europa.eu/legal-content/EN/TXT/PDF/?uri=CELEX:52018DC0237.

European Commission (2019): "Ethics Guidelines for Trustworthy AI, report", https://digital-strategy.ec.europa.eu/en/library/ethics-guidelines-trustworthy-ai.

European Commission (2020): "On Artificial Intelligence. A European Approach to Excellence and Trust, White Paper", COM(2020) 65 final, https://commission.europa.eu/document/download/d2ec4039-c5be-423a-81ef-b9e44e79825b_en?filename=commission-white-paper-artificial-intelligence-feb2020_en.pdf.

Regulation (EU) 2016/679 of the European Parliament and of the Council of 27 April 2016 on the Protection of Natural Persons With Regard to the Processing of Personal Data and on the Free Movement of Such Data, (GDPR).

Directive (EU) 2023/2225 of the European Parliament and of the Council of 18 October 2023 on Credit Agreements for Consumers, (CCD).

Deutscher Bundestag, Drucksache 20/10859 of 27 March 2024, Draft of the First Act to Amend the Federal Data Protection Act (BT- Drucksache 20/10859).

Regulation (EU) 2024/1689 of the European Parliament and of the Council of 13 June 2024 Laying Down Harmonised Rules on Artificial Intelligence, (AI Act).

"Trust" and "Trustworthiness" in the AI Act

Lucia Franke, Benjamin Müller

Abstract *In this paper, we examine the use of "trust" and "trustworthiness" within the AI Act and supporting EU documents on AI. We argue that the underlying concept is limited to reliability, which misconstrues trust as a calculable object and therefore neglects its fundamental meaning. In contrast to this techno-centric view we briefly sketch the idea of trust to demonstrate the essential social and interpersonal aspects of trust, which cannot be neglected in this manner. Our final remarks reflect these considerations within the broader context of the EU's ethical framework.*

> "In a context of rapid technological change, we believe it is essential that trust remains the bedrock of societies, communities, economies and sustainable development. We therefore identify **Trustworthy AI as our foundational ambition**, since human beings and communities will only be able to have confidence in the technology's development and its applications when a clear and comprehensive framework for achieving its trustworthiness is in place." (EG: 4)

In 2024, the Council of the European Union pioneered the world's first comprehensive official AI regulation, titled the Artificial Intelligence Act (AI Act). It may surprise some that one of its key aspects is the stipulation that Artificial Intelligence (AI) shall be "trustworthy". What does this mean? How can AI be trustworthy? And what concept of trust underlies these requirements?

To answer these questions, this paper first examines the use of "*trust*" and "*trustworthiness*" within the AI Act and the supporting Ethics Guidelines, while also briefly outlining the general EU framework. Subsequently, it interprets the conception of trustworthiness found in these texts primarily as reliability and then criticizes it on that basis. Third, it presents an alternative understanding of trust to highlight a perspective that may be missing from the EU framework. Finally, it summarizes the main considerations and offers a brief outlook on possible further steps.

1. The EU's Usage of Trust and Trustworthiness in AI Documents

With the EU AI Act, the EU Commission aimed to achieve two key goals: creating legal certainty for providers and promoting trust[1] among users, including customers and national public administrations, regarding the use of AI. This aim reflects the Act's explicit acknowledgement of "the need to build trust" (Recital 6 AI Act; Impact Assessment Executive Summary: 2). The legislator seeks to achieve this, in part, by promoting "human-centric and trustworthy artificial intelligence" (Art. 1 (1) AI Act; Recital 1 AI Act). Despite this emphasis, the Act itself provides no explicit definition or further explanation of "trustworthy AI". What do "human-centric" and "trustworthy" mean in this context?

Neither the definitions in Art. 3 nor the subsequent normative provisions offer a substantive clarification of these key terms. The recitals suggest a definition rooted in overarching principles (Recital 7 AI Act) and alignment with broader EU values and fundamental rights (see e.g., Recital 27 AI Act), yet they ultimately fall short of providing a precise conceptual framework.[2]

To gain a clearer conceptual understanding of this use of "trust" – or, more precisely, the predominantly used term "trustworthiness" – it is helpful to examine the earlier policy initiatives and guidance documents that have informed EU legislation on artificial intelligence.[3] Among these, the *Ethics Guidelines for Trustworthy AI* (hereinafter "EG"), drafted by the *EU High-Level Expert Group on Artificial Intelligence* (AI

---

1 Interestingly, the noun "trust" itself appears only three times within the Recitals of the AI Act. In contrast, the adjective "trustworthy" is used more frequently.

2 They suggest that trustworthiness depends, at a minimum, on risk mitigation (Recital 64 AI Act), safety, transparency, institutional oversight, and privacy (Recital 68 AI Act), and adherence to ethical AI principles (Recital 165 AI Act).

3 While the European Commission, as early as April 2018, when it presented its plan for 'Artificial Intelligence for Europe,' had already determined that the development and application of AI in the EU required 'an environment of trust and accountability around the development and use of AI', COM(2018) 237 final, p. 13; in order to build and strengthen this trust, measures for data protection and IT security, as well as for the explainability of AI systems – the argument being that those who do not understand how AI works cannot trust it – and through effective legal remedies for those harmed by AI were already considered necessary at that time, COM(2018) 237 final, pp. 14 et seq) , this notion did not translate into concrete regulatory language. In the White Paper on Artificial Intelligence published in February 2020, the Commission presented a comprehensive policy concept for AI regulation, which was based on the two pillars of an 'ecosystem of trust' and an 'ecosystem of excellence,' thereby strongly emphasizing the importance of 'trust,' yet again without providing a clear definition of the term or engaging with the concept in a substantive manner (White Paper on Artificial Intelligence – A European approach to excellence and trust, Brussels, 19.02.2020, COM(2020) 65 final, pp. 1, 3). Only the Impact Assessment, which served as the basis for the initial draft of the AI Act, includes in its glossary the definition of 'Trustworthy AI' derived from the EG.

HLEG) – a body explicitly established by the European Commission – stands out.[4] While not legally binding, these Guidelines were referenced during the preparatory process for the legislative proposal, within the text of the AI Act itself (Art. 95 (2) (a) AI Act) and several times in its recitals (Commission, SWD(2021) 84 final, Part 1/2: 2, 10, 38, 41). They describe their understanding of trustworthy AI in terms of three major components:

> "(1) it should be lawful, ensuring compliance with all applicable laws and regulations, (2) it should be ethical, ensuring adherence to ethical principles and values and (3) it should be robust, both from a technical and social perspective since to ensure that, even with good intentions, AI systems do not cause any unintentional harm. Each component is necessary but not sufficient to achieve Trustworthy AI. Ideally, all three components work in harmony and overlap in their operation. Where tensions arise, we should endeavor to align them." (EG, Conclusion: 35)

To elaborate on these three abstract keywords (*lawful*, *ethical* and *robust*), a comprehensive framework is also established. Proceeding from fundamental rights and four Ethical Principles, the group develops seven key requirements for trustworthy AI: human agency and oversight; technical robustness and safety; privacy and data governance; transparency; diversity, non-discrimination and fairness; societal and environmental well-being and accountability.[5] None of these requirements alone, nor all of them collectively, can ensure trustworthiness; rather, they are intended to form the foundation for possible trust.

In addition, these seven key requirements constitute the foundation of an assessment list for trustworthy AI. This list underwent a piloting process with over 350 stakeholders. The final edition was published on July 17, 2020, with the following promotion: "Through the Assessment List for Trustworthy AI (ALTAI), AI principles are translated into an accessible and dynamic checklist that guides developers and deployers of AI in implementing such principles in practice." (ALTAI) The website also advertises with the slogan: "Measure if your organization's AI is trustworthy". Although the authors of the EG are indeed aware that trustworthiness does

4 The lack of explicit definitions or conceptual clarifications within the AI Act itself is consistent with earlier stages of EU regulation development on artificial intelligence. Already the Commission's initial Communication on "Artificial Intelligence for Europe" (COM(2018)237 final, April 2018) and later the White Paper on Artificial Intelligence (February 2020) broadly emphasized the critical importance of trust without concretely defining or operationalizing the term. The Ethics Guidelines developed by AI HLEG therefore constitute an essential interpretative resource in this respect.

5 The AI Act interprets these as ethical principles. See Recital 27 AI Act. The wording of the EG itself isn't strict, they also refer to this list as principles. See EG: 14, Fn 35.

not arise merely from rules and law, and that it cannot be guaranteed, produced, or even definitively verified. They even remind readers explicitly "that such assessment list will *never be exhaustive*. Ensuring Trustworthy AI is not about ticking boxes, but about continuously identifying requirements, evaluating solutions, and ensuring improved outcomes throughout the AI system's lifecycle, and involving stakeholders therein" (EG: 31).

Thus, it is a complicated and ongoing process for a technology to gain trustworthiness, one with no definitive endpoint. Instead, trustworthy AI requires continuous consideration and implementation within the broader intellectual approach. "Beyond developing a set of rules, ensuring Trustworthy AI requires us to build and maintain an ethical culture and mind-set through public debate, education and practical learning." (EG: 9) These two aspects – cultivating an ethical culture and fostering an appropriate mindset – are central to this endeavor, creating an environment in which trust can grow and new technologies like AI can gain trustworthiness. The recitals of the AI Act and the Ethical Guidelines clearly state that both are based on the fundamental rights of EU Treaties and the EU Charter, which are rooted in human dignity. Therefore, the entire approach of this framework is labeled as "human-centric" (See AI Act R 1, 6 and 27 as well as EG: 9 et seq.).[6]

Concerning the trustworthiness of AI, the overall message conveyed by these documents appears to be that AI is trustworthy if it is embedded in the general framework of the EU, its value system, and its set of beliefs. Instead of defining trust, these texts focus on the necessary circumstances for trust.
Compliance with these framework conditions is thus presented as a mechanism that confers trustworthiness.

2. Which Meaning of Trust?

Having examined the presentation of "Trust" and "Trustworthiness" in the AI Act and the EG, the question of which specific understanding of trust underlies these

6 The glossary of the EG additionally sums up the idea of Human-Centric AI: "The human-centric approach to AI strives to ensure that human values are central to the way in which AI systems are developed, deployed, used and monitored, by ensuring respect for fundamental rights, including those set out in the Treaties of the European Union and Charter of Fundamental Rights of the European Union, all of which are united by reference to a common foundation rooted in respect for human dignity, in which the human being enjoy[s] a unique and inalienable moral status. This also entails consideration of the natural environment and of other living beings that are part of the human ecosystem, as well as a sustainable approach enabling the flourishing of future generations to come." EG: 37.

requirements arises. The AI Act and EG provide no explication of "trust" or "trustworthiness". A simplistic or reductive definition would, in any case, likely undermine the subtle and complex social nature of this phenomenon. Nevertheless, these documents must presuppose a shared, perhaps common, understanding of "trust" and "trustworthiness". The terms themselves, however, allow for a range of interpretations. A framework alone does not guarantee an appropriate understanding of trust. Without further explanation or specification, it remains unclear which particular conception is used. What kind of trust, then, is meant?

Before analyzing the specific understanding of "trust" and "trustworthiness" in the AI Act and EG, it is important to distinguish between these two notions. These terms are too often used interchangeably in general discourse, and, it seems, at times also in the EU documents themselves.

Trust usually refers to a subjective, relational stance or attitude of a person or community towards another entity (be it a person, institution, or system). It involves vulnerability, expectation, and unpredictability.

Trustworthiness, conversely, refers to the objective qualities, characteristics, or adherence to norms possessed by an entity that make it worthy of being trusted. It is a property independent of whether anyone actually decides to trust.

While these notions are connected, neither is a necessary or sufficient condition for each other. When someone is trustworthy, it does not mean that everybody automatically trusts him. Conversely, one can be trusted yet not be trustworthy at all. Worthiness alone is no guarantee for anything; yet, it is still a desirable quality.

The analysis in the following subsections will argue that the EU framework primarily focuses on establishing conditions for AI's trustworthiness, largely interpreting this concept through the lens of technical reliability and compliance (see Section 2.1). It is possible that the AI Act, at least in parts, uses the term "trustworthiness" in ways specific to IT security law (see Recital 27 CRA) or a political agenda (e.g., Ribeiro et al., 2016, among others).[7] The incorporation of technical "*trustworthiness levels*" for risk categorization and certification into the Act demonstrates the partial integration of such a technical understanding of trust into the regulatory framework.[8]

---

7 Recital 123 AI Act, which stipulates that a conformity assessment is required to ensure a high level of trustworthiness, can be interpreted as meaning that trustworthiness is achieved solely through this assessment, which itself incorporates the concept of trustworthiness levels from Regulation 2018/881.

8 According to ENISA, cybersecurity and trustworthiness are closely intertwined: "the requirements of trustworthiness complement and sometimes overlap with those of AI cybersecurity in ensuring proper functioning. [...] Hence, trustworthiness features such as robustness, oversight, accuracy, traceability, explainability and transparency inherently support and complement cybersecurity" (ENISA, *Cybersecurity of AI and Standardisation*, 2023: 9–10). However, ENISA also points out that trustworthiness depends not only on cybersecurity, but also on regulatory oversight and control (ENISA, *Cybersecurity of AI and Standardisation*, 2023: 23).

Furthermore, "trustworthy AI" serves as a political objective and guiding slogan prominent in the policy discourse preceding the formal AI Act proposal. Foundational documents, such as the European Commission's White Paper on Artificial Intelligence, emphasized "trust" as a prerequisite for public acceptance and the successful economic development and deployment of AI technologies within the Union (Proposal: 1, 30; White Paper: 1, 3). In this context, *"trustworthiness"* was framed as an economic necessity – based on the rationale that only trustworthy AI would ultimately succeed in the market (Proposal: 30) – rather than as a precisely defined technical or ethical attribute.

2.1. Trustworthiness as Reliability

Nevertheless, the underlying concept of trust in the EU documents seems to be orientated towards economical and mechanical modes of thought, which filter through the wording of the texts under consideration, even though the practice and culture of trust are also emphasized.

The recitals of the AI Act already allude to a functional, measure-based understanding of trustworthiness (e.g. Recital 64, 123 AI Act). References to specific measures, such as mandatory risk management systems, conformity assessments, and technical solutions like content watermarking, exemplify the measure-based dimension; the functional aspect is highlighted through principles implying operational characteristics like technical robustness and transparency, and by tying requirements to the system's defined purpose and context (Laux et. al. 2024: 3, 6; otherwise Ho and Caals 2024: 358).

Robustness is also one of the three key components for trustworthy AI in the EG. The EG elaborate on the need for robust AI as follows:

> "Even if an ethical purpose is ensured, individuals and society must also be confident that AI systems will not cause any unintentional harm. Such systems should perform in a safe, secure and reliable manner, and safeguards should be foreseen to prevent any unintended adverse impacts. It is therefore important to ensure that AI systems are robust. This is needed both from a technical perspective (ensuring the system's technical robustness as appropriate in a given context, such as the application domain or life cycle phase), and from a social perspective (in due consideration of the context and environment in which the system operates). Eth-

---

ENISA thus posits that the AI Act's requirements (data and data governance, record-keeping, transparency and provision of information to users, human oversight, risk management system, quality management, conformity assessment, robustness) aim to establish a trustworthy AI ecosystem, although they often coincide with cybersecurity requirements (ENISA, *Cybersecurity of AI and Standardisation*, 2023: 19–20).

> ical and robust AI are hence closely intertwined and complement each other." (EG: 7)

This quotation provides significant insight into the underlying understanding of trustworthiness and is merely one example among others. Systems should be safe, secure, and reliable, potential harm needs to be prevented. However, the call to "*prevent any unintended adverse impacts*" is the antithesis of trust: It is a call for control. Meanwhile, robustness itself is a technical term, arguably ill-suited and imprecise for social issues. What "trustworthy" appears to mean here is *reliability* (see also Costa 2024: 39).

Accordingly, the other two components – "*lawful*" and "*ethical*" – would then simply mean reliably conforming to law and ethics. This interpretation is also supported by the phrasing in the EG's conclusion: AI "should be lawful, ensuring compliance with all applicable laws and regulations, (2) it should be ethical, ensuring adherence to ethical principles and values" (EG, p. 35). Compliance and adherence leave no room for deviation and thus no opportunity to genuinely trust.

Reliability is indeed a core component of trustworthiness. An entity that consistently fails to perform as expected – such as a person who regularly breaks promises or a doctor whose diagnoses are mostly wrong – is justifiably deemed not trustworthy. However, while reliability is central to trustworthiness, the fundamental issue with reducing the concept of trustworthiness solely to technical reliability, or with equating trustworthiness (even if encompassing reliability) with trust itself, is the neglect of the freedom aspect and the relational, often uncertain, dynamics inherent in genuine trust. Trust is not simply inherent in an entity's demonstrated reliability or trustworthiness. The main issue with limiting trust or trustworthiness to reliability is the neglect of the necessary aspect of freedom in trust and, consequently, of all social and interpersonal meanings of trust. Doing so makes trust or trustworthiness seem like a calculable object, one that can be measured, operationalized, or even produced, as the assessment list in the Guidelines implies (EG: 24 et seq. and ALTAI).

Unlike technical reliability (and robustness), a person's trust cannot be engineered, calculated, verified, manufactured, or produced. It does not even have a definitive status. Trust is a special relation between two free beings, possessing its own rules rather than adhering to formal laws. It always remains vague and uncertain and is arguably never robust by any means.

The Guidelines acknowledge in their glossary that "trust is usually not a property ascribed to machines" (p. 38). However, the authors of the Guidelines evidently have no problem with calling AI systems potentially trustworthy; furthermore, they emphasize the need to ensure that these systems – along with all associated individuals and processes – are trustworthy.

As Kaminski (in this volume) emphasize, genuine trust in technological artifacts has always been critically dependent on the institutional frameworks and contexts that create, regulate, oversee, and assume responsibility for these technologies. Institutionalized trust emerges from mechanisms of accountability, transparency, certification, liability structures, political accountability, and ongoing societal negotiation processes. Thus, when we speak of trustworthy AI, we are often referring to the trustworthiness of the institutional and human processes surrounding AI – those that develop, deploy, regulate, monitor, and take responsibility for AI systems (Ho and Caals, 2024: 368; Costa 2024: 37). To meaningfully speak of trustworthy AI, therefore, political and regulatory frameworks must move beyond mere technical standards.

2.2. A problem of expertise?

A possible explanation for this strong emphasis on technical reliability in the understanding of Trustworthy AI could lie in the composition of the expert group. To borrow an insight from the group itself: One potential problem might be the lack of diverse skills and competencies in the team developing ethical guidelines (cf. EG: 25, see also pp. 18, 23). Although "ethics" is in the title (and is described in its own glossary as a philosophical discipline), the "high level expert group" includes only three professional philosophers among its 52 members.[9] To be fair, It is indeed an expert group on AI. But why should an expert on AI also be an expert in AI ethics? Without questioning their expertise on AI, one can question their expertise in ethics and their awareness of the subtle distinctions and challenges inherent in human thought and social practices, including trust. For this subject, perhaps a different group composition would have been more appropriate.

Another indication of this issue is that the EG's definition of "Trust" in its glossary is simply taken "from the literature" (EG: 38) – a business paper by an information scientist and his student, rather than from any humanities scholar – and it does not substantially address ethical aspects. This borrowed definition, however, seems more appropriate than the conception of trust developed in the main body of the EG. Amusingly, the cited paper itself states: "Ethics and governance of artificial intelligence are areas that need more attention." (Siau and Wang 2018: 52) Ironically, subsequent attention was then drawn to the very paper that had called for it.

None of this is necessarily a problem, although it is, at a minimum, questionable. On the one hand, there is an issue of expertise; on the other, potentially stemming

9 These are Coeckelbergh, Floridi and Metzinger. You get six professional philosophers if you add one ethicist (Van Wynsberghe) and two legal scholars, who also work on Philosophy of law (Hilgendorf and Yeung).

from this, there may be a deficient understanding of the phenomenon of trust itself, particularly when it is reduced to reliability.

2.3 Structural Deficits of the AI Act: Insufficient Context Sensitivity and Top-Down Approach

These conceptual shortcomings, notably the inclination to equate trustworthiness with reliability (as discussed in Section 2.1), may also be reflected in the AI Act's structural approach, which struggles with the inherent context-dependency of trustworthiness (the focus of this section). The inherent context-dependency of trustworthiness challenges attempts to systematically enumerate conditions for "*trustworthy AI*". The AI Act contains a relatively rigid evaluative framework in which obligations significantly rely on a static risk classification (Laux et. al. 2024: 7; Nasr-Azadani 2024: 19). While this classification primarily addresses potential harms, this focus on harm prevention aligns only partially with the broader measures needed to establish trustworthiness. Consequently, this framework conflicts with the dynamic nature of trust, limiting adaptive, context-sensitive assessment (in essence, the measures dictated by the harm-centric classification may significantly exceed the requirements for trustworthiness in certain cases while potentially falling short in others).

Underlying this structural rigidity is the Act's predominantly top-down approach to defining and enforcing trustworthiness (exception in case of "*special trust domains*", Ho and Caals, 2024: 360). Specifically, the AI Act's legislative approach primarily views trustworthiness as a top-down condition to be imposed externally, based on adherence to prescribed criteria, rather than as a socially negotiated and collectively emergent phenomenon (Laux et al. 2024: 4, 7). While these criteria may not be explicitly intended as the sole guarantee of trust, this approach poses the risk of "*compliance theatre*" in practice – that is, focusing on actions that create the appearance of meeting regulatory requirements without necessarily achieving substantive goals, such as genuine trustworthiness (Costa 2024: 39). Companies deploying or providing AI might primarily view the regulatory requirements as a checklist to be completed, leading to the misleading assumption that formal compliance equates to being fully trustworthy. By emphasizing measurable criteria over process and potentially fostering this "*tick-box*" mentality, the AI Act may overlook or insufficiently recognize the social dimension of trust-building. Consequently, it risks omitting or even hindering processes essential for authentic trust formation.

3. The idea of trust

Following the critique of current approaches, it seems helpful to sketch, at least in broad strokes, the idea of trust. This perspective is offered not as a comprehensive trust theory but rather as an indication of what might be missing when "trustworthiness" in contexts like the EU AI Act and EG is predominantly framed in technical terms and limited to reliability.

Trusting always involves the freedom of another person and the inherent risk that accompanies it. When we trust, we, in a sense, expect the unexpected: We hold a positive expectation about an outcome that we ultimately cannot fully control, precisely because it relies on free action. Consider trusting a friend with a secret: We depend on their discretion despite their freedom to choose otherwise. Trust becomes essential precisely where direct control ends – whether because we are overwhelmed by the situation or because other persons are involved, whose actions are also free and therefore always pose a certain risk; one can never be sure what another person will do next. Freedom, ironically, is the prerequisite for trust and, at the same time, the reason it is needed in the first place. Speaking of trust is meaningful only in connection with freedom. Without a free being involved, there is no need for trust (and, in a strict sense, it is not even possible).

Beyond this lies the miracle of trust: that it can provide actual certainty regarding the unpredictable. "When I trust someone, his certainty of himself is for me the certainty of myself; I recognize my own Being-for-myself in him, I recognize that he recognizes my Being-for-myself and that it is for him purpose and essence." (Hegel 2018: 219). Trusting means that I am certain that my purpose and essence are also his. This marvellous connection of trust would require an extensive examination.

This idea contrasts sharply with our relations to traditional machines, which operate deterministically according to their design or programming. It makes little sense to speak of "trusting" them in an interpersonal way. This distinction arguably extends to so-called artificial intelligence, which is neither free nor genuinely intelligent; it operates statistically and simply performs calculations based on algorithms (Laux et al. 2024: 4). This fundamental characterization applies across the spectrum, from large language models to other types of machine learning models or systems, even if their specific capabilities and how we interact with them differ markedly.

Considering trust through this lens uncovers another paradox in the quest for "*trustworthy AI*": Systems are required to be more predictable and reliable (see "*Reliability and Reproducibility*," EG: 17) to be deemed trustworthy – a demand that runs counter to the acceptance of freedom and unpredictability inherent in the general idea of trust.

Perhaps the common shift in discourse from "trusting AI" to "trustworthy AI" implicitly acknowledges this contradiction. The question is then no longer: "*Can you trust AI?*" but is, instead, reversed: "*Is the AI at least worthy of trust?*" The concept of trustworthiness avoids the critical difficulty of directly trusting an AI or other machine. Instead, it describes the relation between a free person and a non-free object, whose properties (such as reliability and safety – meaning that it can be used without posing a danger) are assessed.

In many cases, especially with complex AI systems, users or even regulators cannot independently verify the system's reliability or adherence to all specified requirements. Therefore, forming a judgment about an AI system's actual reliability or compliance usually necessitates placing trust in others – in the developers, certifiers, regulators, or third-party auditors who conduct assessments and provide assurances. This introduces an often-overlooked layer: Establishing trust in the AI system's operation is not merely a matter of verifying its properties but frequently requires prior trust in the institutions and individuals evaluating and overseeing that system's reliability and compliance. Trust is not a technical problem but a social issue.

This sketch also raises a concluding philosophical question regarding regulatory efforts in general: Can the quality that invites authentic trust truly be established or guaranteed through regulation and technical compliance alone? Or is such an endeavor inherently contradictory, risking a fundamental misunderstanding of trust itself? While ensuring AI systems are reliable and safe is undeniably crucial, this alternative perspective suggests that equating technical "trustworthiness" with a richer, freedom-based understanding of trust may be insufficient.

4. Final remarks

In any case, a deeper understanding is still needed of what kind of tool AI actually is, how to use it responsibly, and for what specific purposes. Simultaneously, we must elaborate on the circumstances under which its developers can be genuinely trusted or deemed trustworthy.

If central concepts remain abstract and disconnected from concrete contexts, there is a danger that they will degenerate into empty keywords or buzzwords in debates. This risk is heightened when such a keyword, like "trust", is prominently highlighted, as seen in the EU's AI Act. Unfortunately, even the best concepts can be misused (one might recall the French Revolution's descent into the Reign of Terror).

Drawing from the analysis in the preceding sections, the EU's approach to trustworthiness appears problematic. As discussed, the term "trust" seems either to lack a distinct, clearly articulated concept – leaving "trust" open to abuse, misleading practices, and contributions to chaotic, inefficient discussions – or to be implicitly limited to mere technical reliability (Laux et al. 2024: 4). This second possibility, potentially suggested by assessment lists implying that trust can be calculated or even produced, is itself misleading (as indicated, for example, by the promotion of the assessment list).

Thus, both potential paths are problematic and challenge the EU's own ethical framework. Even if the concept of trust cannot be defined with absolute precision, it must be related to everyday practices and challenges. The analysis presented here underscores the need for a more profound engagement with the underlying conception of trust. The framework claims to be human-centric, yet the concept of trust it employs seems predominantly tech-centric, lacking social dimensions such as freedom and vulnerability that typically accompany issues of trust (Costa 2024: 31).

Understanding the intended concept of trust requires familiarity with the broader framework, not just the AI Act alone. While embedding AI in general frameworks is the right direction, further steps are clearly needed. To effectively cultivate trust and move beyond purely technical solutions, a comprehensive governance strategy is necessary. This strategy must move beyond the technical perspective of reliability (as discussed in Section 2.1) and purely regulatory top-down mandates (as discussed in Section 2.3) and incorporate the relational and social dimensions of trust outlined in Section 3. Such a strategy could incorporate elements such as institutionalizing mechanisms for transparent ethical deliberation to address complex, domain-specific value conflicts; establishing clear accountability structures through defined institutional responsibilities and robust liability regulations; or implementing continuous monitoring processes to track real-world performance and impacts, thereby enabling dynamic and adaptive trust. Despite such potential practical steps, the entire issue demands deeper conceptual thought – which, along with political considerations, is fundamentally a task for philosophy and law.

References

Costa, Maria I. (2024): "Building on the EU's Unique Strategy for Artificial Intelligence (AI). Can an Ethical Foundation Be Successfully Integrated into Its Design and Deployment?", in: UNIO – EU Law Journal, 10(1), pp. 30–41.

Hegel, G. W. F. (2018): The Phenomenology of Spirit, Translated by M. Inwood, Oxford University Press.

Ho, Calvin W. L. and Caals, Karel (2024): "How the EU AI Act Seeks to Establish an Epistemic Environment of Trust", in: Asian Bioethics Review, 14(1), pp. 345–372.

Laux, Johann, Wachter, Sandra and Mittelstadt, Brent (2024): "Trustworthy Artificial Intelligence and the European Union AI Act. On the Conflation of Trustworthiness and Acceptability of Risk", in: Regulation & Governance, 18(1), pp. 3–32.

Nasr-Azadani, Mohamad. M. and Chatelain, Jean. L. (2024): "The Journey to Trustworthy AI-Part 1. Pursuit of Pragmatic Frameworks, arXiv:2403.15457.

Ribeiro, Marco T., Singh, Sameer and Guestrin, Carlos (2016): "Why Should I Trust You? Explaining the Predictions of Any Classifier, in: Proceedings of the 22nd ACM SIGKDD International Conference on Knowledge Discovery and Data Mining, pp. 1135–1144.

Siau, Keng and Wang, Weiyu (2018): "Building Trust in Artificial Intelligence, Machine Learning, and Robotics", in: Cutter Business Technology Journal, 31(7), pp. 6–13.

Legal Sources

Assessment List for Trustworthy Artificial Intelligence (ALTAI) for self-assessment (2020), https://digital-strategy.ec.europa.eu/en/library/assessment-list-trustworthy-artificial-intelligence-altai-self-assessment, last access: July 21, 2025.

Commission Staff Working Document Executive Summary of the Impact Assessment Report (2021): Brussels, 21.4.2021, SWD(2021) 85 final, 2021/0106(COD). [Impact Assessment Executive Summary]

Commission staff working document impact assessment accompanying the proposal for a regulation of the European Parliament and of the Council laying down harmonised rules on artificial intelligence (Artificial Intelligence Act) and amending certain Union legislative acts (2021), SWD/2021/84 final, Part 1. [Proposal]

Commission (2020), White Paper on Artificial Intelligence – A European approach to excellence and trust, Brussels, 19.02.2020, COM (2020) 65 final, pp. 1, 3. [White Paper]

High-Level Expert Group on Artificial Intelligence (2019): Ethics Guidelines for Trustworthy AI [EG]

Assessment List for Trustworthy Artificial Intelligence (ALTAI) for self-assessment (2020), https://digital-strategy.ec.europa.eu/en/library/assessment-list-trustworthy-artificial-intelligence-altai-self-assessment, last access: July 21, 2025.

Regulation (EU) 2024/1689 of the European Parliament and of the Council of 13 June 2024 laying down harmonised rules on artificial intelligence and amending Regulations (EC) No 300/2008, (EU) No 167/2013, (EU) No 168/2013, (EU) 2018/858, (EU) 2018/1139 and (EU) 2019/2144 and Directives 2014/90/EU, (EU) 2016/797 and (EU) 2020/1828 (Artificial Intelligence Act)

Regulation (EU) 2024/2847 of the European Parliament and of the Council of 23 October 2024 on horizontal cybersecurity requirements for products with digital elements and amending Regulations (EU) No 168/2013 and (EU) 2019/1020 and Directive (EU) 2020/1828 (Cyber Resilience Act)

Appendix

Author Profiles

Sebastian Clemens Bartsch is a Ph.D. candidate at the Chair of Information Systems and Electronic Services at the Technical University of Darmstadt, Germany. He holds a Master's degree in Business Information Systems from the Technical University of Darmstadt. His research interests include accountability in the context of AI-based information systems and explainable AI.

Oliver Behn is a PhD student at the Chair of Digitalization and Process Management at the University of Marburg and a research associate at the Chair of Agricultural and Food Business Management at the University of Göttingen. His research interests focus on behavioral spillover effects, behavior change, and technology adoption.

Alexander Benlian is a professor of information systems at TU Darmstadt, Germany. His research focuses on algorithmic management, accountable artificial intelligence, IT entrepreneurship, and human-computer interaction. His work has been published in journals such as MIS Quarterly, Information Systems Research, and Entrepreneurship Theory & Practice. He serves as senior editor at European Journal of Information Systems, associate editor at Information Systems Research, and as department editor at Business & Information Systems Engineering. He also serves on the editorial boards of MIS Quarterly Executive and Electronic Markets.

Michael Birkner, LL.M. (University of Glasgow), is a doctoral candidate at the Institute for Law and Regulation of Digitalisation (IRDi) at Marburg University, Germany and former research assistant of the chair of Professor Möslein.

Gregor Bosold is a doctoral candidate and former member of the chair of Professor Dubovitskaya.

Sebastian Bücker works as a scientific researcher at the Institute of Philosophy, TU Darmstadt. Using both his backgrounds in computer science (M.Sc.) and philosophy of technology (M.A.), his research focusses on how current developments in AI can be connected to philosophy, especially regarding the ascription of capabilities associated with autonomy.

Elena Dubovitskaya is a professor of Civil Law, Commercial and Corporate Law and Digitalization Law at the Justus Liebig University in Giessen. Her research focuses on the interconnections between Law and AI, especially the explainability of AI for legal purposes.

Marcus Düwell is professor of philosophy at the Institute for Philosophy at the Technical University of Darmstadt (Germany) after positions in Tübingen (Germany) and Utrecht (the Netherlands). His research interests include foundational questions of moral and political philosophy, philosophical anthropology, and applied ethics. Publications entail the Cambridge Handbook on Human Dignity (2013) and (together with Deryck Beyleveld) The Sole Fact of Pure Reason. Kant's Quasi-Ontological Argument for the Categorical Imperative (2020).

Nico Formánek is head of the Philosophy of Computational Sciences department at the High-Performance Computing Center Stuttgart. He is interested in the intersection of computing technology, science and society and has worked on the epistemology of computer models in physics, the problem of induction in machine learning and sustainability in high performance computing.

Lucia Franke, Dr. jur., B.A., is a research associate at the Chair for Civil Law, Business Law and Banking Law held by Prof. Dr. Katja Langenbucher at Goethe University Frankfurt. She completed her doctorate in legal argumentation theory, her research focuses on civil law, civil procedure law, the law of digitalization (with an emphasis on data protection and AI), as well as legal philosophy and legal theory.

Marc Jungtäubl is a researcher in organizational sociology at FernUniversität in Hagen and at Hessische Hochschule für öffentliches Management und Sicherheit. His research areas include digital formalization of work, AI ethics in organizations, and governance frameworks for AI trustworthiness assessment. Marc Jungtäubl examines how professionals navigate AI systems in healthcare settings while maintaining autonomy and competence. He has published extensively on the formalization of work, human-AI interaction, and the humane design of work and working conditions.

Andreas Kaminski is a professor of philosophy of science and technology at the Technical University of Darmstadt. His research areas include social epistemology (philosophy of trust and testimony), technical epistemology (the role of technology in science) and politics of technology (modeling for policy). Andreas Kaminski is co-editor of the Yearbook of Philosophy of Technology. He is a senior scientist at the High-Performance Computing Center of the University of Stuttgart (HLRS).

Katja Langenbucher is a law professor at Goethe-University, affiliated professor at SciencesPo, Paris, and visiting faculty at Fordham Law School, NYC. She has held visiting positions at Sorbonne; WU Vienna; LSE; Columbia, Fordham and PennLaw School and will join NYU Law Global Faculty in 2026. She holds supervisory position as BaFin and SciencesPo.

Michael Leyer, PhD, is a professor of business administration having the chair of Digitalization and Process Management at the University of Marburg as well as being adjunct professor at Queensland University of Technology. The main focus of his research is the sociotechnical consideration of future technologies in organizations.

Florian Möslein, Dr. iur., Dipl.-Kfm., LL.M. (London), is Director of the Institute for Law and Regulation of Digitalisation (www.irdi.institute) and Professor of Law at the Philipps-University Marburg, where he teaches Contract Law, Company Law and Capital Markets Law. He previously held academic positions at the Universities of Berlin, St. Gallen, and Bremen, and visiting fellowships in Italy (Florence, European University Institute), the US (NYU, Stanford and Berkeley), Australia (University of Sydney), Spain (CEU San Pablo, Madrid) and Denmark (Aarhus). Having graduated from the Faculty of Law in Munich, he also holds academic degrees from the University of Paris-Assas (licence en droit) and London (LL.M. in International Business Law). Florian Möslein published three monographs and over 80 articles and book contributions, and has edited seven books. His current research focus is on regulatory theory, corporate sustainability and the legal challenges of the digital age.

Benjamin Müller, M. A., is a research associate in the ZEVEDI project group "Digital Governance. Responsibility and Trust in Digital Structures (DigiGov)". He is pursuing his doctorate in philosophy about freedom, based on Hegel and Schelling. Along with Steffen Augsberg and Marcus Düwell he co-edited a volume regarding data access rules ("Datenzugangsregeln") in 2024.

Sebastian Omlor, Dr. iur., LL.M. (NYU), LL.M. Eur. (University of Saarland), is a Full Professor and Director of the Institute for Law and Regulation of Digitalisation (IRDi) at Marburg University, Germany. He holds the chair for Private Law, Commercial and Business Law, Banking Law and Comparative Law. Beside the field of international commercial law, his research focuses banking and company law, the law of payment and financial services, the law of digitalisation (www.irdi.institute), and the legal concept of money.

Philipp Richter is a professor of philosophy at the Institute of Philosophy I, Department of Philosophy and Educational Sciences at Ruhr-University Bochum, Germany. Since 2019 he has held the Chair of Didactics of Philosophy and Practical Philosophy. His research is focused on normative ethics and applied ethics, and on Didactics of Philosophy, the research about teaching and learning of philosophy.

Jan-Hendrik Schmidt is a Research Associate at the Chair of Information Systems and Electronic Services at the Technical University of Darmstadt in Germany, where he also earned his Ph.D. in Information Systems Management. His research explores how artificial intelligence is integrated into information systems, with a focus on accountability and responsibility. He also investigates how advances in AI shape the development of information systems, using both quantitative and qualitative research methods.

Hans Wilke, LL.M. (Columbia), is a research assistant and doctoral candidate at the Institute for Law and Regulation of Digitalisation (IRDi) at Marburg University, Germany.

Mascha Will-Zocholl is Professor of Work and Organisational Sociology at the Hessian University of Applied Sciences for Public Management and Security (Wiesbaden). Prior she worked at Goethe-University and Technische Universität Darmstadt. Her research focuses on the informatization and digitalisation of work and administration. She investigates how digital technologies transform work processes, professional identities, skills, organisations and the organisation of work in a spatial perspective. She is co-editor of the Lexicon of Work and Industrial Sociology (2021) and the anthology "Topologies of Work. How Digitalisation and Virtualisation Shape Working Spaces and Places" (together with Caroline Roth-Ebner), published by Palgrave Macmillan in 2021. She wrote about 'Information Space(s)' in "Workplace Theories. A Handbook of Theories on Designing Alignment Between People and the Office Environment" (edited by Rianne Appel-Meulenbroek and Vitalija Danivska).

PI13578033
9718739